Dr. Todd Zimmerman
Madison Area Technical College

Student Workbook to Accompany

INTRODUCTORY PHYSICS: BUILDING UNDERSTANDING

Jerold Touger
Curry College

JOHN WILEY & SONS, INC.

Cover Photo Credit: © Stuart Westmorland/Image Bank/Getty Images

To order books or for customer service please, call 1-800-CALL WILEY (225-5945).

ISBN-13 978-0-471-68311-7
ISBN-10 0-471-68311-6

Printed in the United States of America

10 9 8 7 6 5 4 3 2 1

Printed and bound by Bind Rite Graphics

Contents

Preface

The skills students develop as they complete the workbook exercises will aid them in solving the more quantitative problems in the textbook.

The workbook exercises lead students through the steps of the scientific reasoning process.

Students will learn to interpret the various representations used, such as sketches, graphs, formulas, and verbal descriptions, and translate a problem between the representations.

The exercises draw on many years of research of the most common conceptual difficulties students encounter.

Name ______________________ Date __________ Class __________

Chapter 1

Physics, Mathematics, and the Real World

1. Describe how each of the following jobs would require knowledge of physics.

   a. Air Force fighter pilot

   b. Meteorologist

   c. Telephone communications technician

   d. Civil engineer

2. Come up with two other careers that would involve knowledge of physics.

3. What units would be most appropriate for measuring the following items? Use the Conversion Factors chart at the front of the text to help you.

   a. The wing span of a large passenger jet.

   b. The length of a football field.

   c. The size of the gears in a wrist watch.

   d. The length of a nail.

e. The thickness of a single strand of human hair.

f. The frequency of a kid on a swing.

g. The mass of a car.

h. Air pressure at sea level.

i. The volume of water in a swimming pool.

j. The volume of water in a coffee mug.

k. The mass of a strand of human hair.

4. Rewrite the answer on the right side of the equation with the appropriate number of significant figures.

   31.6 in = 2.6333 ft

   ________ft

   433 h = 18.0417 days

   _______days

   1.00 km = 0.6214 mi

   ________mi

   11.1 lb = 49.373 N

   _______ N

   2.64 ft = 0.804633 m

   ________m

3.7 L = 0.130647 ft^3

________ft^3

1.602×10^{-19} J = 1.0000000 eV

________eV

5280 ft = 1.609 km

________km

100.0 ft = 1200 in

________in

38 mi/h = 61.142 km/h

________km/h

2.3 ft × 7.183 ft = 16.5209 ft^2

________ft^2

0.83 m × 0.367 m × 0.22 m = 67.014 L

________L

5. Covert the numbers to the indicated units. Be sure to answer with the appropriate number of significant figures.

1.00 mi = ________km

0.90 mi = ________km

1.0 ft = ________m

3.98 ft = ________m

1.0 cm = ________in

6.254 cm = ________in

1.00 days = ________yr

Name ____ **Date** _______________ **Class** _______________

1 m/s = ________ft/s

3.08 m/s = ________ft/s

31.3 yr = ________s

71.30 km = ________mi

Name ______________________ Date ____________ Class ____________

Chapter 2

Describing Motion in One Dimension

1. The picture is a snapshot of a particular instant.

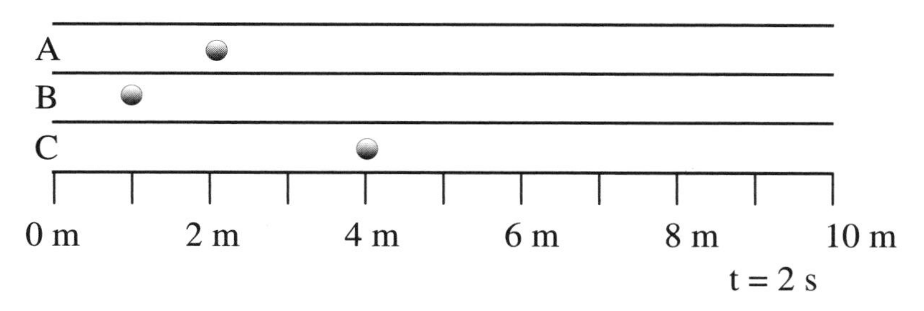

a. Which of the following pieces of information can you get from the snapshot: position of a ball, displacement of the ball, a time interval?

A snapshot at a later instance shows the balls have moved.

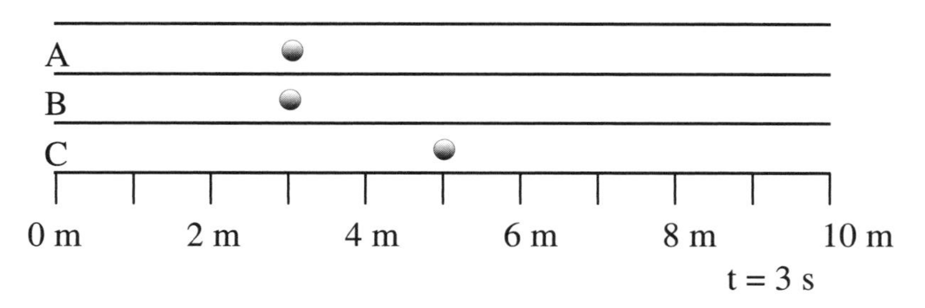

b. Which of the following pieces of information can you get from the two snapshots: position of a ball, displacement of the ball, a time interval?

c. Which ball is moving the fastest? How do you know which is moving fastest?

Name ________________________________ Date ____________ Class ____________

d. Calculate the displacement for each ball.

e. What is the time interval between the two pictures?

f. Calculate the average velocity for each ball.

g. If balls A, B, and C are moving at a constant speed, sketch the location of each ball for the next three seconds.

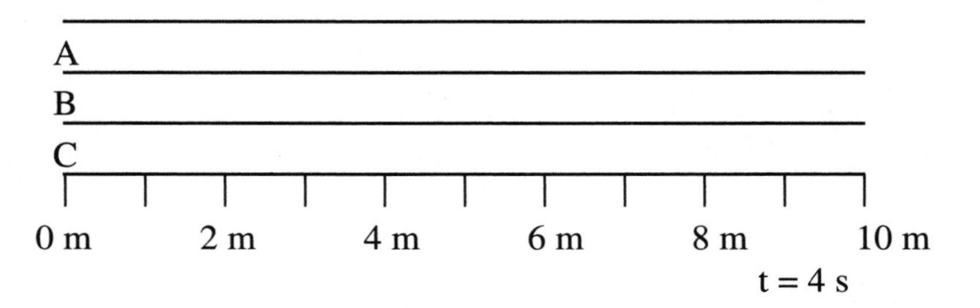

2. The picture below shows the location of three balls moving at different speeds at one second intervals. You can think of the picture like a multiple exposure using a strobe light that flashes once each second. The number below each ball is the clock reading in seconds.

A
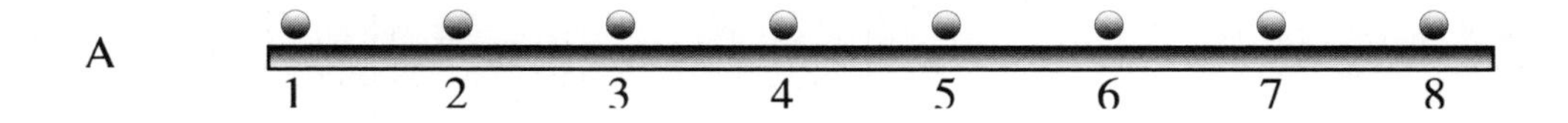

B
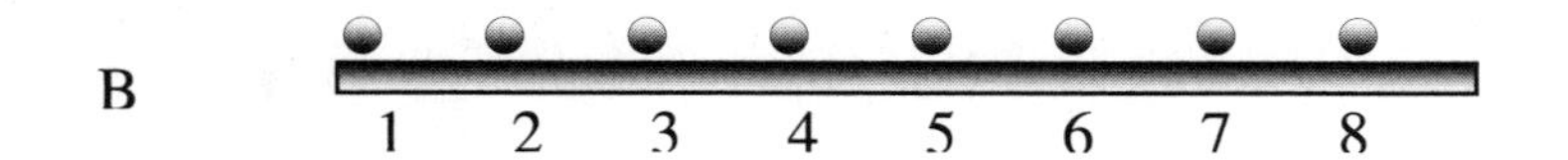

C
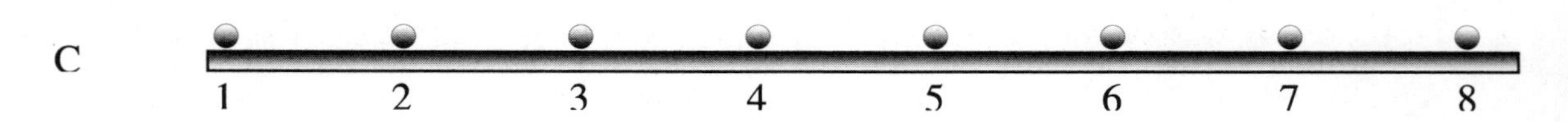

a. Which ball is moving the fastest? How did you determine that?

Name ______________________________ Date ____________ Class ____________

b. Which ball is moving the slowest? Explain your reasoning.

c. Use a ruler to measure how far each ball moves between t = 4 s and t = 6 s.

d. The time interval between the two points is $\Delta t = 6\text{ s} - 4\text{ s} = 2\text{ s}$. Calculate the average speed of each ball.

e. Measure how far each ball moves between the first and last point.

f. Calculate the time interval between the first and last point.

g. Calculate the average speed for each ball.

3. Two balls, one with constant velocity and one accelerating, are shown in the strobe picture below. The clock readings in s are shown. Each mark is 1 m apart.

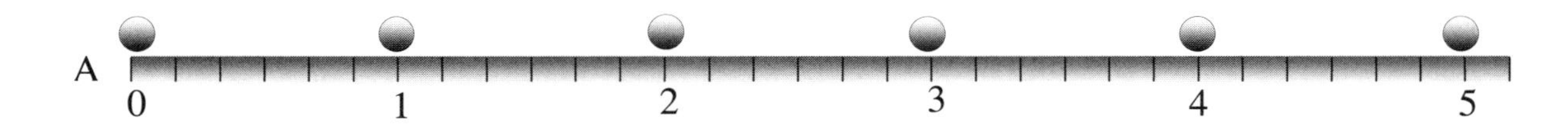

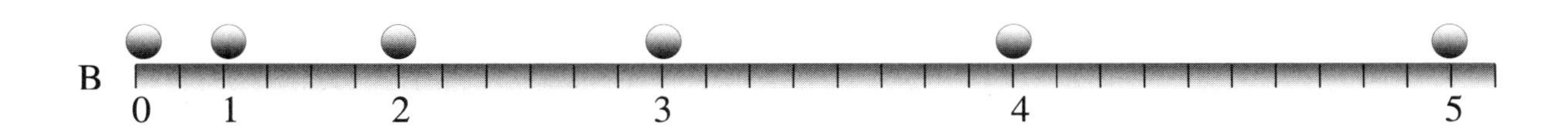

a. Without doing any calculations, which ball has a larger average velocity for the first three seconds? Which ball has a larger average velocity for the entire run?

b. Calculate the average velocity for A and B for the first three seconds.

c. Find the displacement for both balls from start to finish. Using this displacement, calculate the average velocity for A and B for the entire time interval. Why do they both have the same average velocity?

d. Is the instantaneous velocity ever equal to the average velocity for ball A? Explain.

e. Is the instantaneous velocity ever equal to the average velocity for ball B? Explain.

f. Is there a time interval during which the two balls have the same velocity?

g. During which time interval do the two balls pass one another?

h. By how much does the velocity of ball A change during each 1 s interval?

i. Calculate the average acceleration for ball A.

j. By how much does the velocity of ball B change during each 1 s interval?

Name ________________________________ Date ____________ Class ____________

k. Calculate the average acceleration for ball B.

4. The strobe picture below shows ball A moving with constant velocity. The numbers to the left of the balls are the clock readings in seconds. The strobe picture has been turned sideways to aid in drawing the graph.

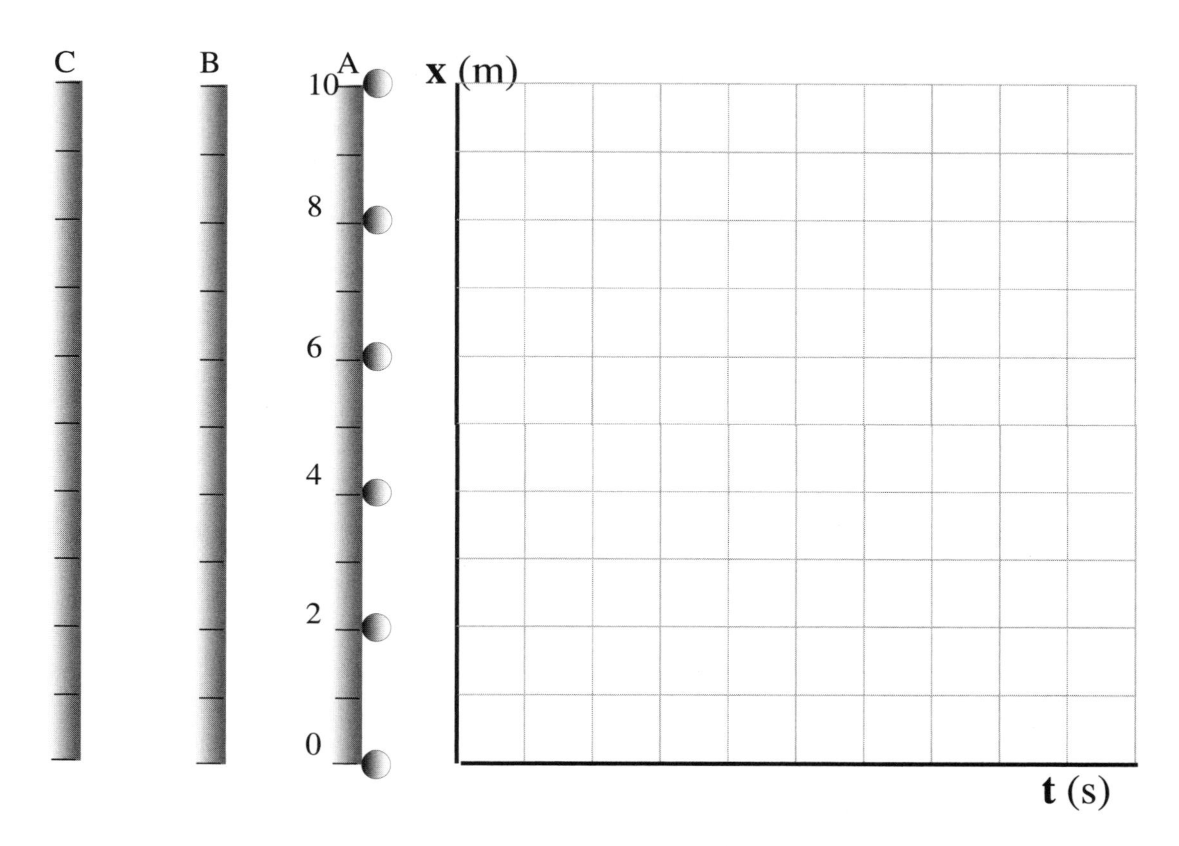

a. Draw a dot on the graph for each position of the ball. Draw a line connecting the points.

b. Draw a strobe picture on the left at B for a ball moving at half the speed.

c. Use a different color to sketch the graph for a ball moving at half the speed.

d. How would the slope for a faster moving ball compare to the two lines already drawn on the graph?

e. Use the line on the far left to draw the strobe pictures of a ball starting at 10 m and moving in the opposite direction but at the same speed as ball A.

Name ______________________ Date ____________ Class ____________

f. Sketch the graph for ball C.

g. What does the steepness of the line tell you about the ball?

h. What information about the movement of the ball does the direction of the slope tell you?

i. How would the graph look if the ball started at 0 m and gradually sped up? In the space below sketch the rough shape of the graph and explain your reasons.

5. The picture below shows a ball speeding up.

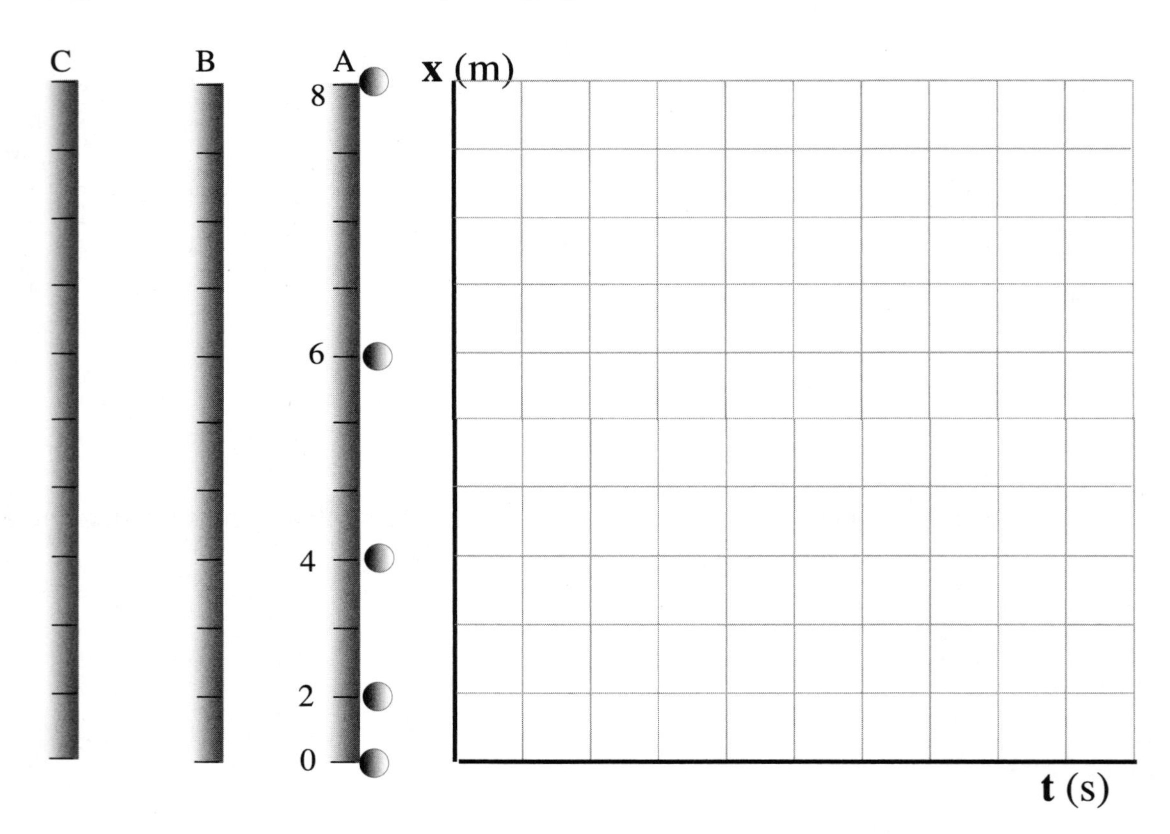

Name ______________________ Date ____________ Class ____________

a. Draw the graph for the ball. Explain why your prediction from the previous question differed from the graph.

b. By how much is the velocity of the ball changing during each interval?

c. How long is each interval above? Use this and your answer to the previous question to calculate the average acceleration.

d. Assuming the ball's initial velocity is 0 m/s at t = 0 s, use the acceleration you just calculated to determine how fast the ball is moving after 10 s.

e. How would the graph look if the ball was initially moving fast but slowing down? Sketch the overall shape of the graph below and explain your reasons.

f. Predict the shape of the graph if the ball was speeding up but in the opposite direction. Sketch the graph and explain your reasons.

Name ________________________________ Date ____________ Class ____________

g. Draw the strobe picture for ball B slowing down. Starting at 0 m, the ball moves 4 m in the first 2 s. The ball travels 1 m less during each of the following time intervals.

h. By how much is the velocity changing during each time interval?

i. Calculate the acceleration of the ball. Pay close attention to the sign of the acceleration.

j. Sketch the graph for ball B. How did your prediction differ from the graph?

k. Draw the strobe picture for ball C speeding up but moving in the opposite direction. Starting at 10 m, the ball moves an additional 1 m each interval.

l. Calculate the acceleration of the ball. Pay close attention to the sign of the acceleration.

m. Sketch the graph for ball C. How did your prediction differ from the graph?

Name ______________________ Date ____________ Class ____________

6. A friend of yours is off to take a physics quiz and asks you for a crash course in motion graphs. Explain to your friend how to tell how fast an object moves, whether it is speeding up or slowing down, and which way it is going from the position versus clock time graphs.

7. You are doing a lab where a motion sensor records your movements towards or away from the sensor. For the set of walking directions, draw a rough sketch of a possible position versus clock reading graph. The motion sensor is located at 0 m.

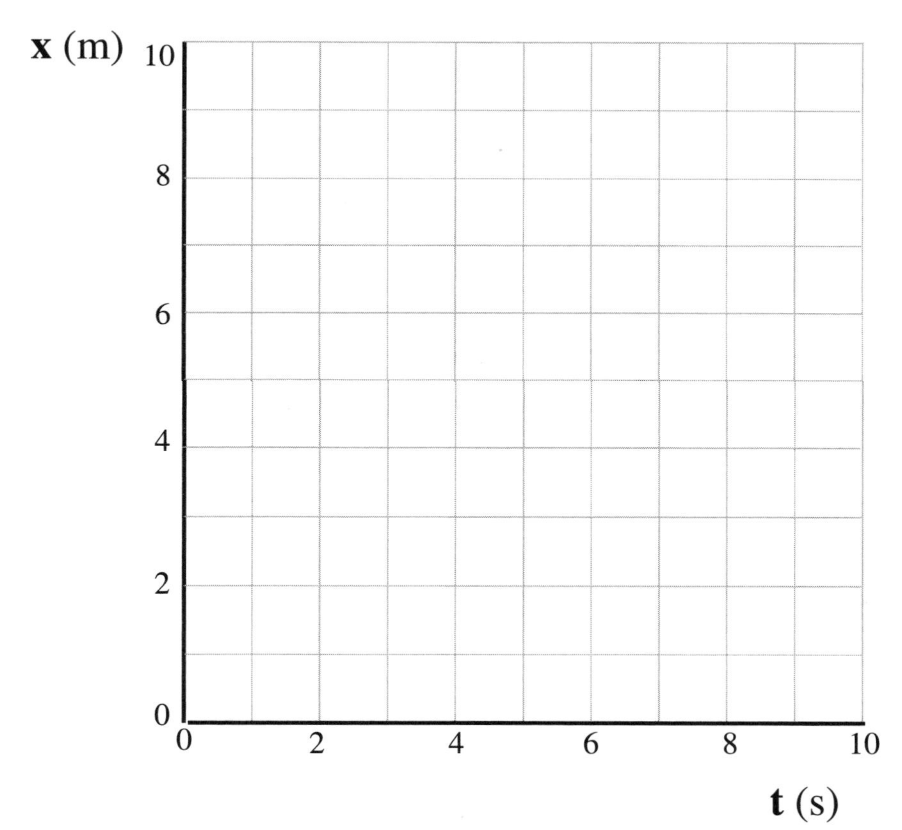

a. Starting right at the sensor, walk away from the sensor slowly for the count of 2.
b. Stop for the count of 2.
c. Walk away from the sensor even slower for the count of 3.
d. Walk back toward the sensor quickly until you reach the sensor.

Name ______________________________ Date ____________ Class ____________

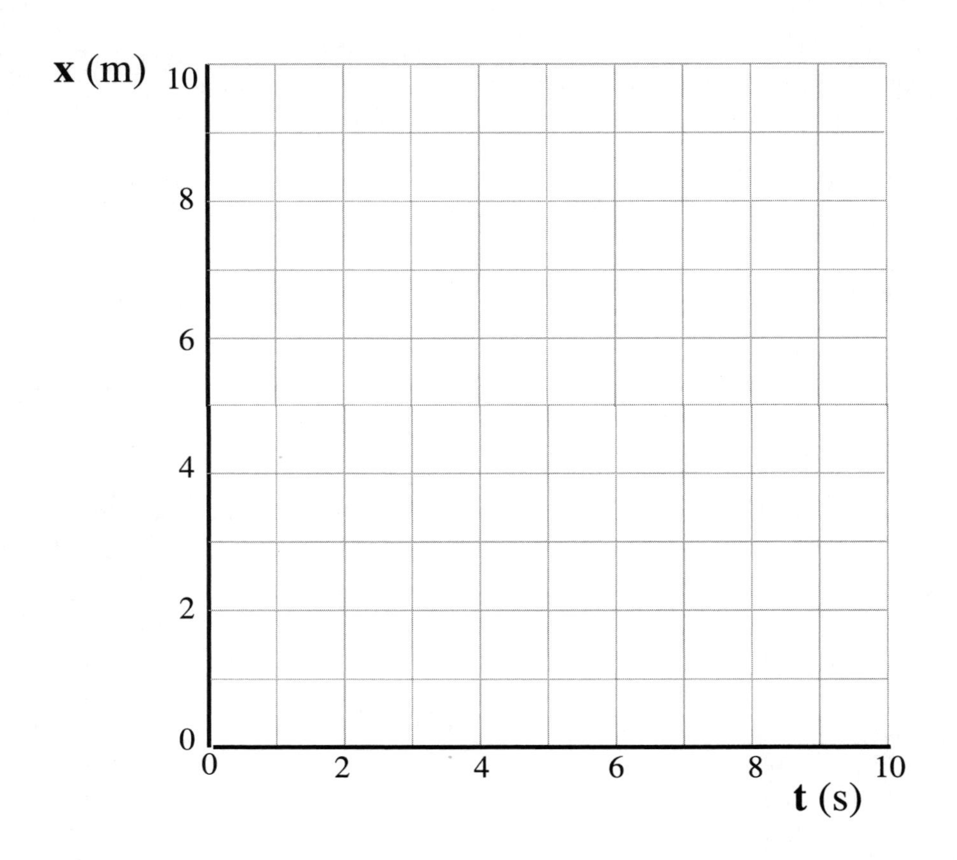

e. Starting 3 m in front of the sensor, quickly walk away from the sensor for 2 s.
f. Walk back toward the sensor at the same speed for 2 s.
g. Walk away from the sensor, slowly at first but with increasing speed for 5 s.
h. Stop for 1 s.

Name ______________________________ Date ____________ Class ____________

8. Write walking directions for the position versus time graphs. Exact distances and speeds are not important.

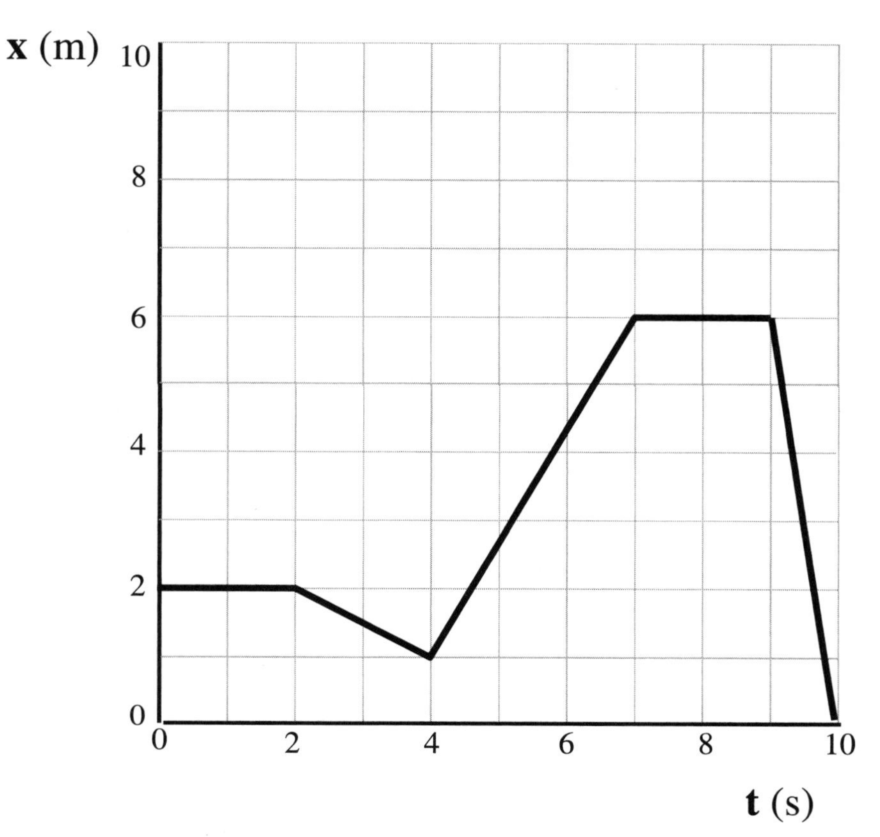

Walking Directions:

Name ______________________ **Date** __________ **Class** __________

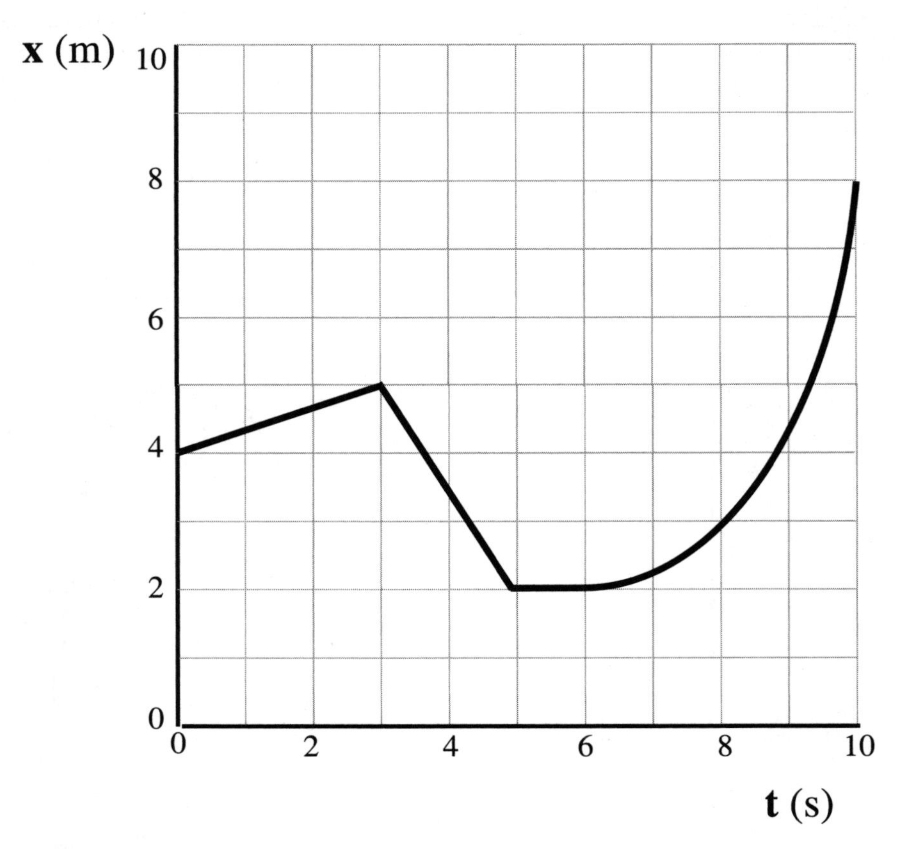

Walking Directions:

Name ______________________________ Date ____________ Class ____________

9. Calculate the average velocity for each two second interval. Plot these values on the velocity versus clock reading graph.

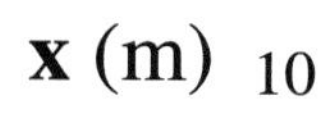

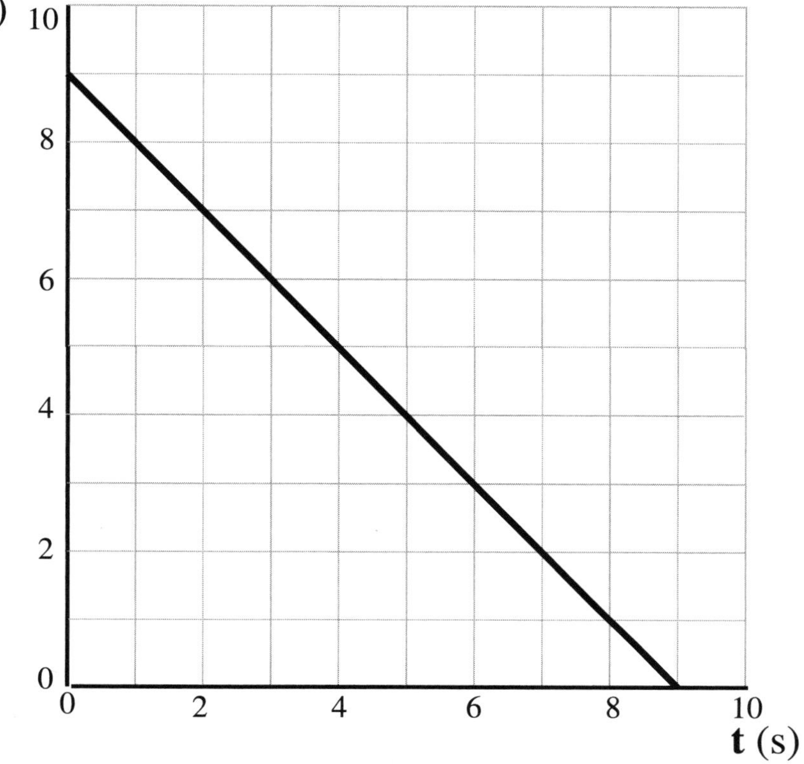

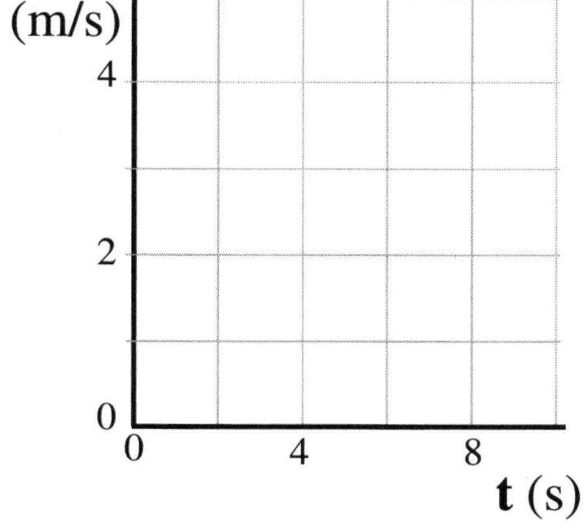

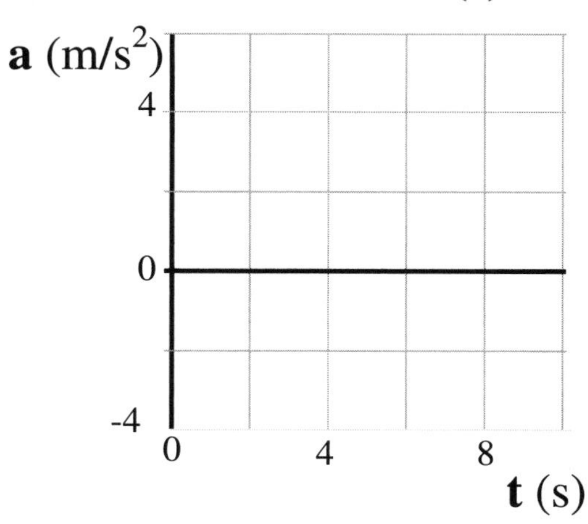

10. Is the object moving at a constant speed, speeding up, or slowing down? How can you tell?

Name ______________________________ Date ____________ Class ____________

11. Calculate how much the velocity changes during each 2 s interval.

12. Calculate the average velocity and plot it on the acceleration versus clock reading graph.

13. In the space below, sketch what you think the velocity versus clock reading graph looks like for an object speeding up.

Name ______________________ **Date** __________ **Class** __________

14. Calculate the average velocity for each two second interval. Plot these values on the velocity versus clock reading graph.

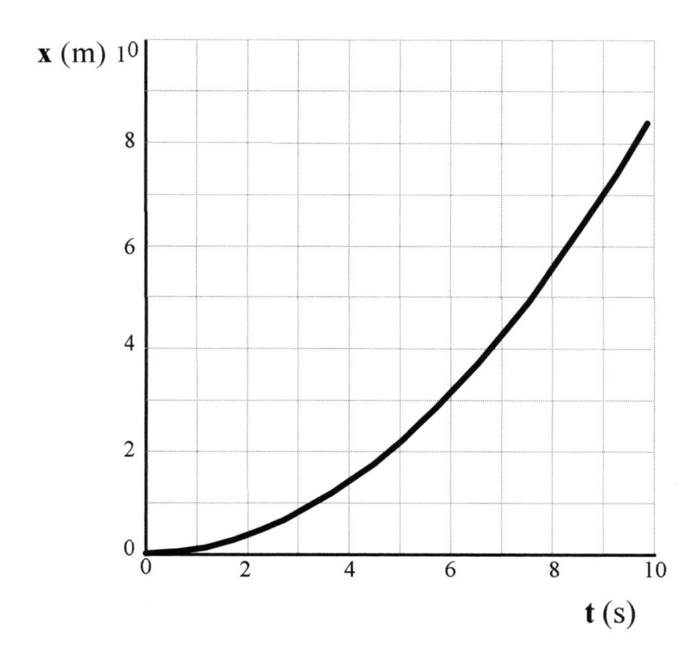

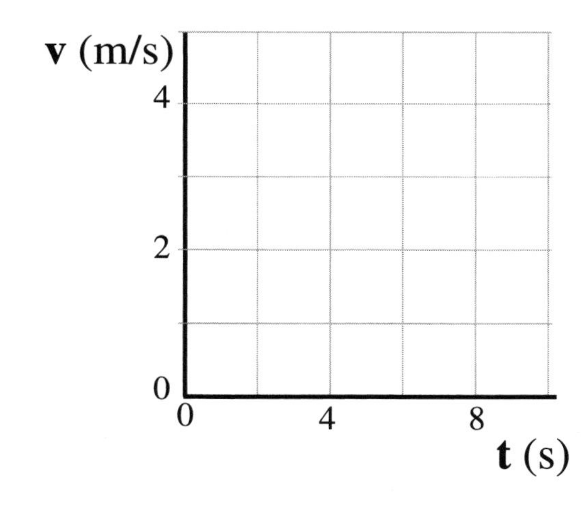

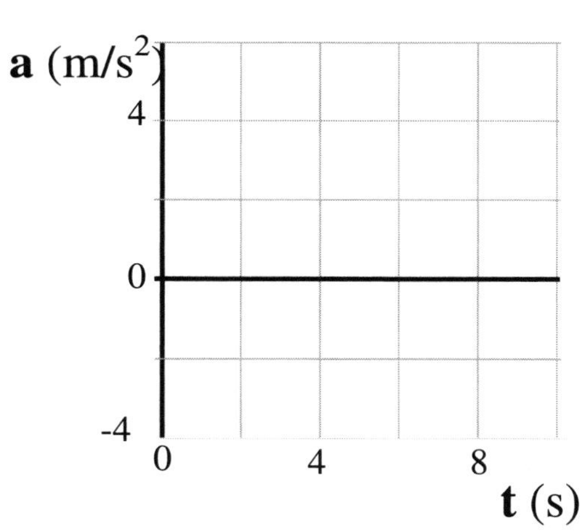

Name ______________________________ Date ____________ Class ____________

15. Is the object moving at a constant speed, speeding up, or slowing down? How can you tell?

16. Calculate how much the velocity changes during each 2 s interval.

17. Calculate the average acceleration and plot it on the acceleration versus clock reading graph.

18. In the space below, sketch what you think the velocity versus clock reading graph looks like for an object slowing down.

Name ______________________________ Date ____________ Class ____________

19. Calculate the average velocity for each two second interval. Plot these values on the velocity versus clock reading graph.

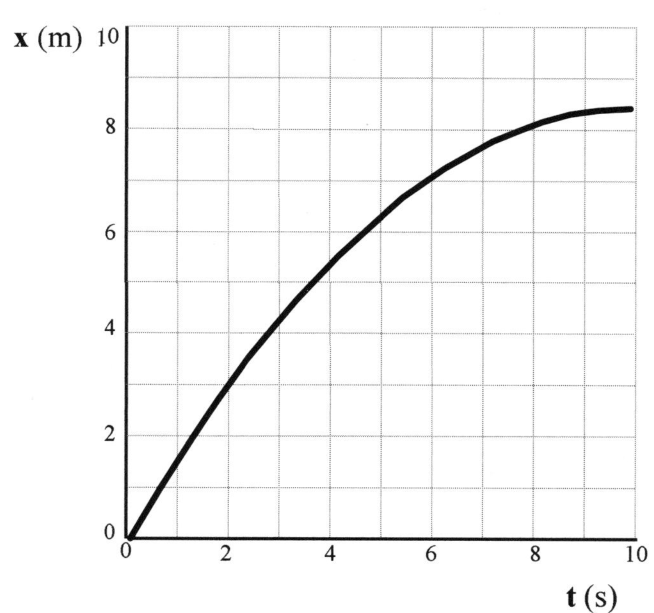

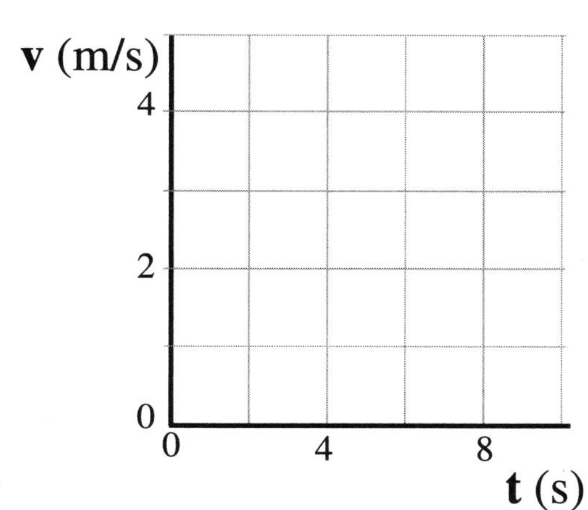

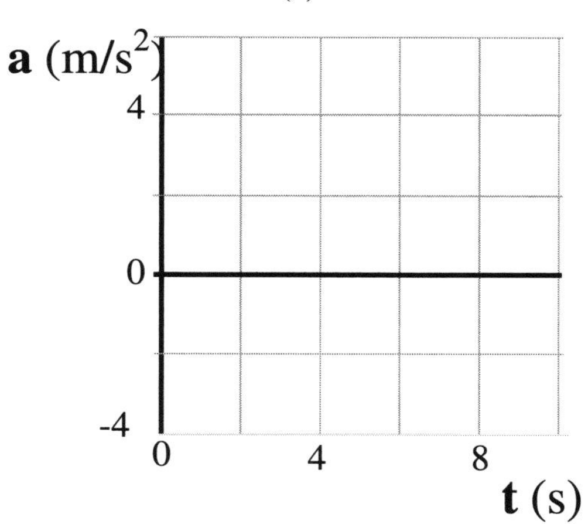

20. Is the object moving at a constant speed, speeding up, or slowing down? How can you tell?

21. Calculate how much the velocity changes during each 2 s interval.

Name ______________________________ Date ____________ Class ____________

22. Calculate the average acceleration and plot it on the acceleration versus clock reading graph.

23. Which of the following information can you calculate from a position versus clock reading graph? Explain how you would find each piece of information: starting position, starting velocity, position at any time, velocity at any time, acceleration. all but acce

24. Which of the following information can you calculate from a velocity versus clock reading graph? Explain how you would find each piece of information: starting position, starting velocity, position at any time, velocity at any time, acceleration.

not statig postion

25. Which of the following information can you calculate from an acceleration versus clock reading graph? Explain how you would find each piece of information: starting position, starting velocity, position at any time, velocity at any time, acceleration.

$a = \frac{V}{t}$

Name ______________________ Date __________ Class __________

26. Use the velocity versus clock reading graph to draw the position versus clock reading graph for an object initially at 2 m.

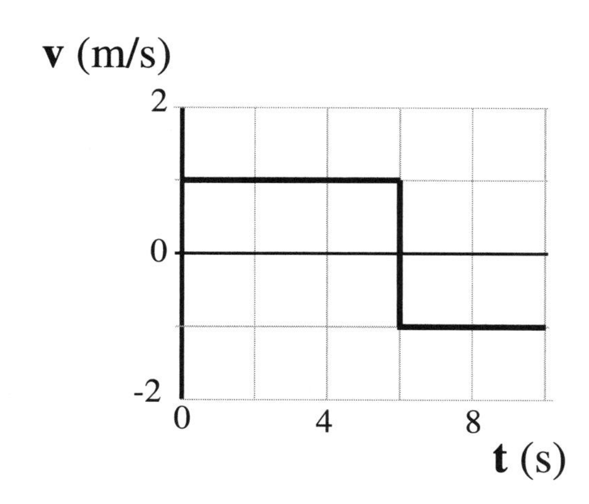

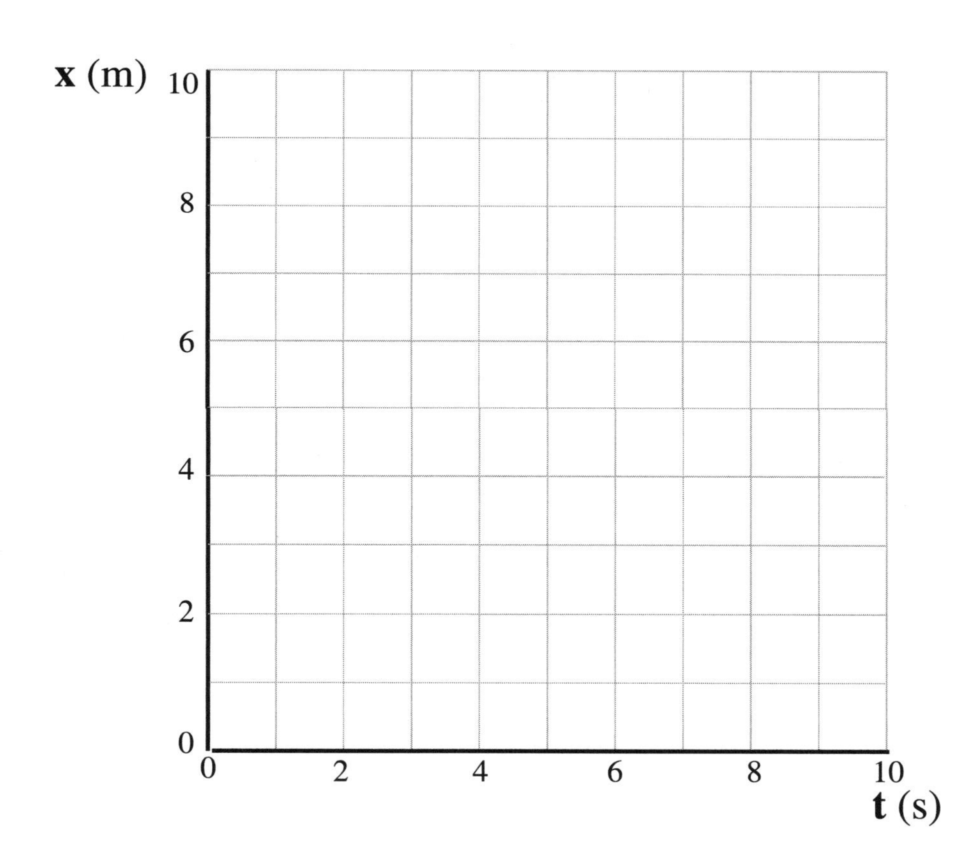

Name ______________________________ Date ____________ Class ____________

27. Use the velocity versus clock reading graph to draw the position versus clock reading graph for an object initially at 3 m.

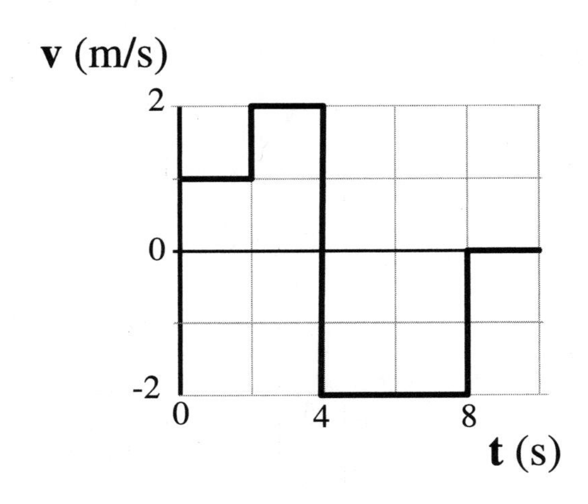

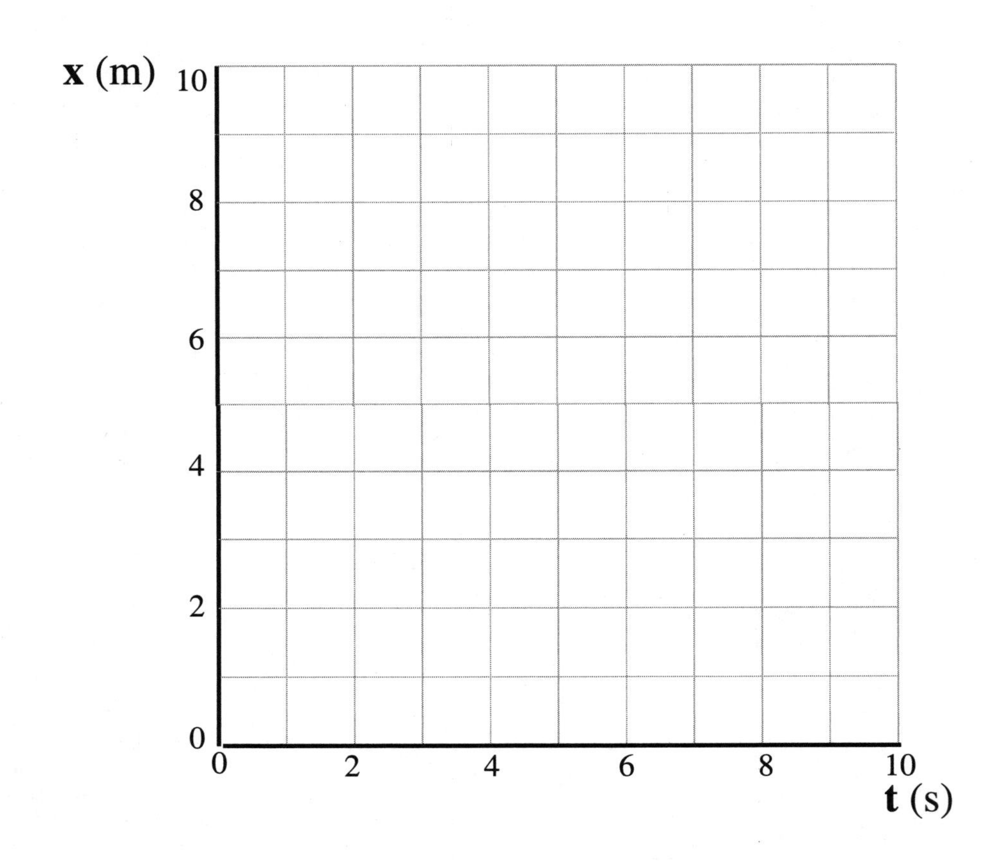

Name ______________________ Date ____________ Class ____________

28. A ball is starting at 0 m, moving 1 m/s and accelerating at 0.5 m/s2. Sketch the strobe picture, velocity versus clock reading and position versus clock reading graphs.

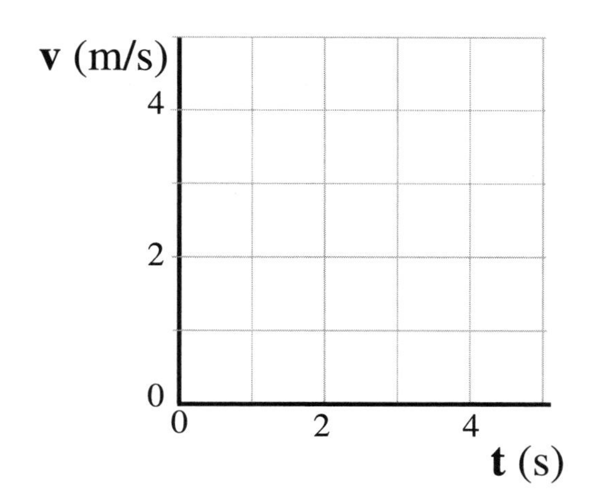

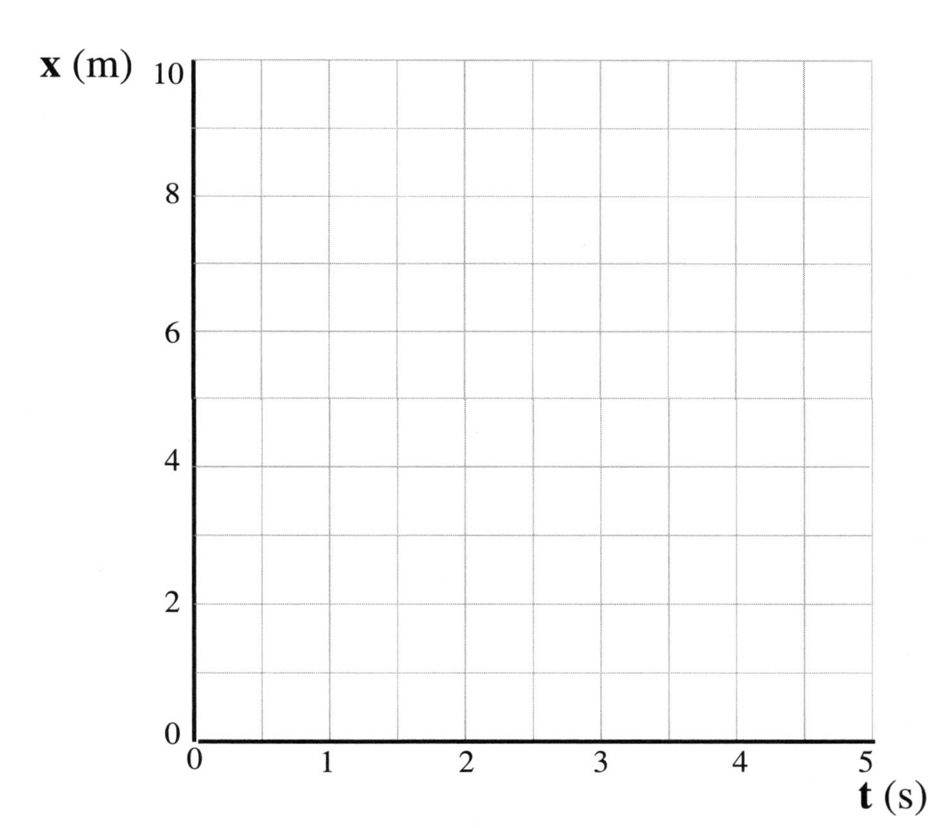

29. When the velocity and acceleration of an object have opposite signs, is the speed of the object increasing or decreasing? Explain.

Name ______________________________ Date ____________ Class ____________

Chapter 3

Constructing Two-Dimensional Motion from One-Dimensional Motions

Vectors

1. Draw the resultant vector **R** = **A** + **B**.

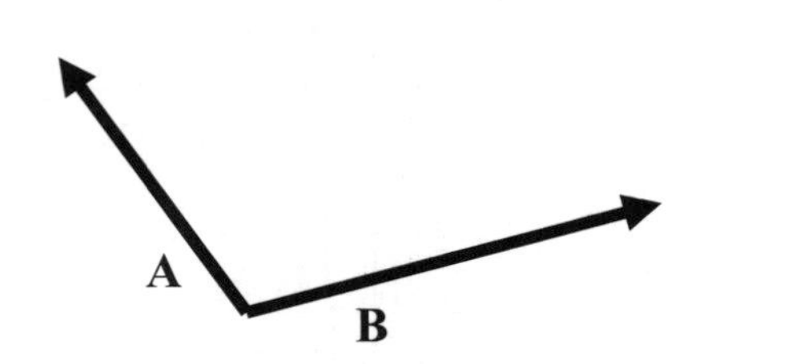

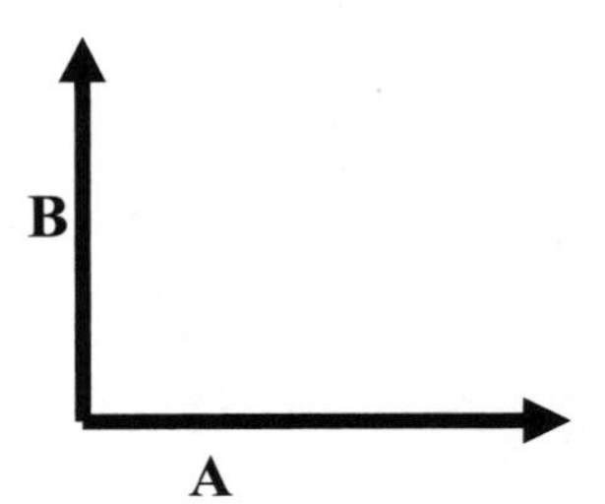

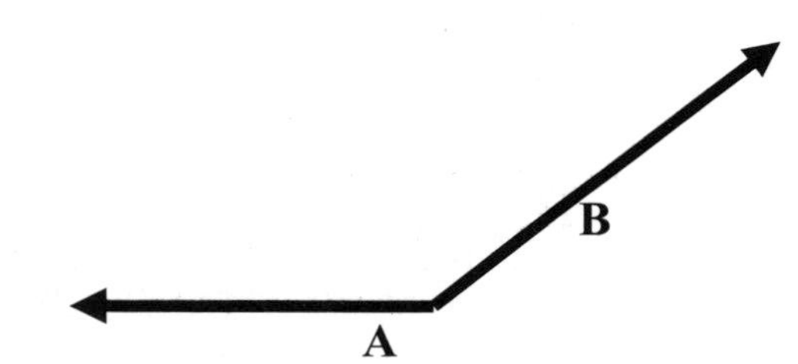

Name ______________________ Date ____________ Class ____________

2. Use a protractor and a ruler to draw the vectors **A** and **B** and the resultant vector **R** = **A** + **B** in the space below.

   a. **A** is 5 cm long at an angle of 30° and **B** is 3 cm long at an angle of 65°.

   b. **A** is 7.5 cm long at an angle of 15° and **B** is 6 cm long at an angle of 180°.

   c. **A** is 6.5 cm long at an angle of 0° and **B** is 6.5 cm long at an angle of 90°.

Name ______________________ Date ____________ Class ____________

3. Use a protractor and ruler to find the angle the vector makes with the x-axis and the length of the vector. Project the scalar components onto the x-axis and y-axis and measure each component. (You have a head start for the first one.)

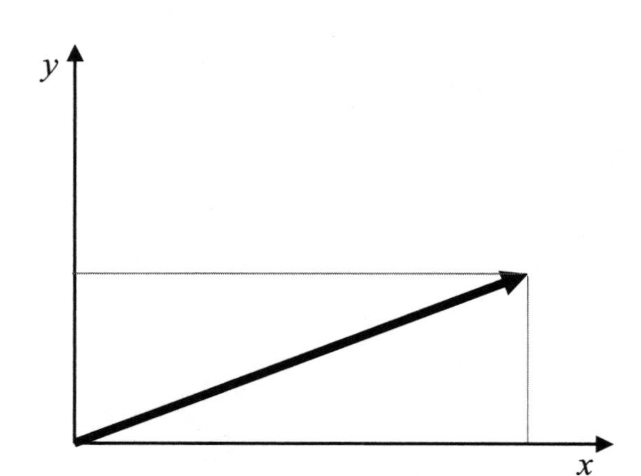

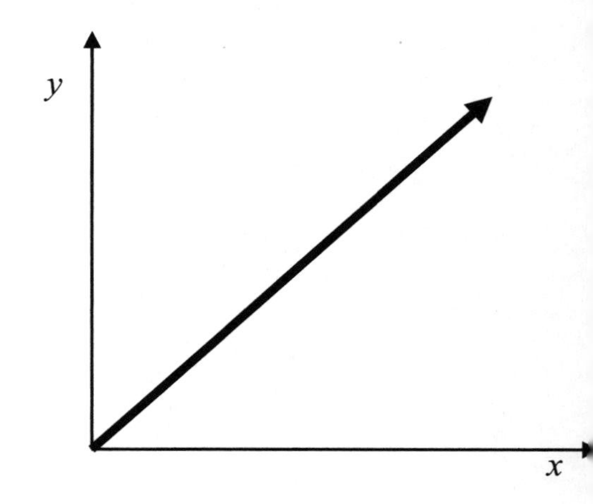

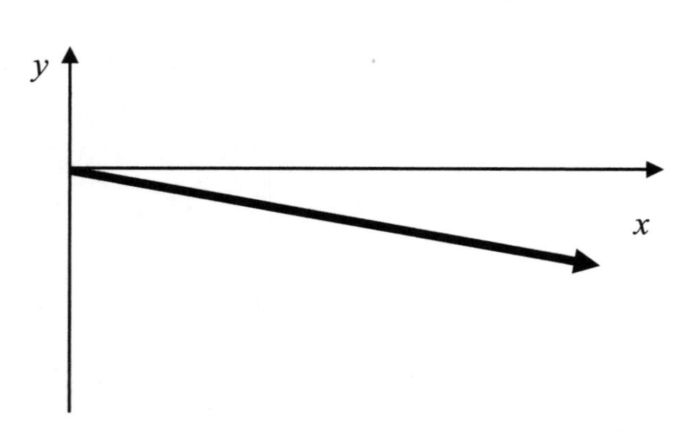

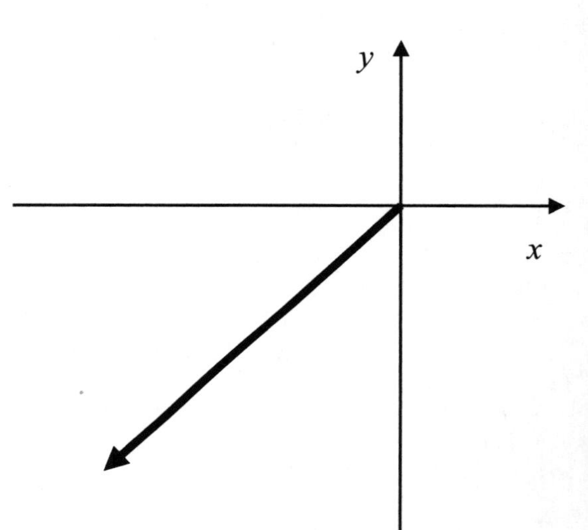

Name ______________________________ Date ____________ Class ____________

4. For each of the three vectors below, complete these steps.

   a. Use a protractor to measure the angle the vector makes with the x-axis and use a ruler to measure the length of the vector.
   b. Draw the x- and y-components on the x- and y-axis.
   c. Calculate the length of each component.
   d. Measure the length of each component.
   e. Compare your measurement to the calculated value.

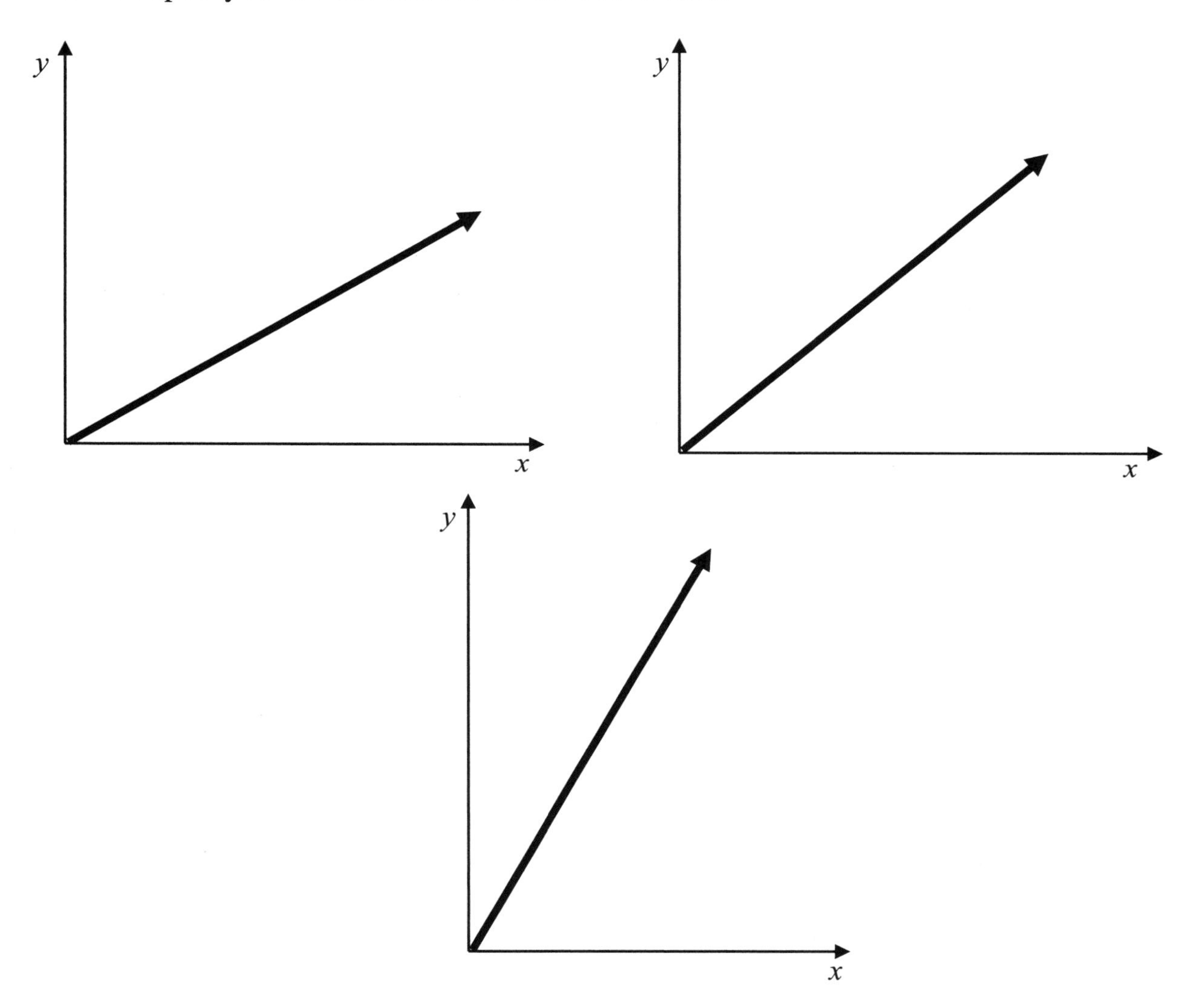

Name ______________________________ Date ____________ Class ____________

5. Draw the components of the vector in the space below. Measure the length of the vector as well as the angle. Calculate the length and angle from the components and compare to the measured values.

   a. x-component = 3.5 cm, y-component = 7 cm

   b. x-component = 8 cm, y-component = 6 cm

Name ______________________________ **Date** ____________ **Class** ____________

6. Calculate the components of the vectors **A** and **B** and add the components together to find the resultant vector **R**. Draw the resultant vector **R**. Graphically add **A** and **B** and compare to the resultant found using components.

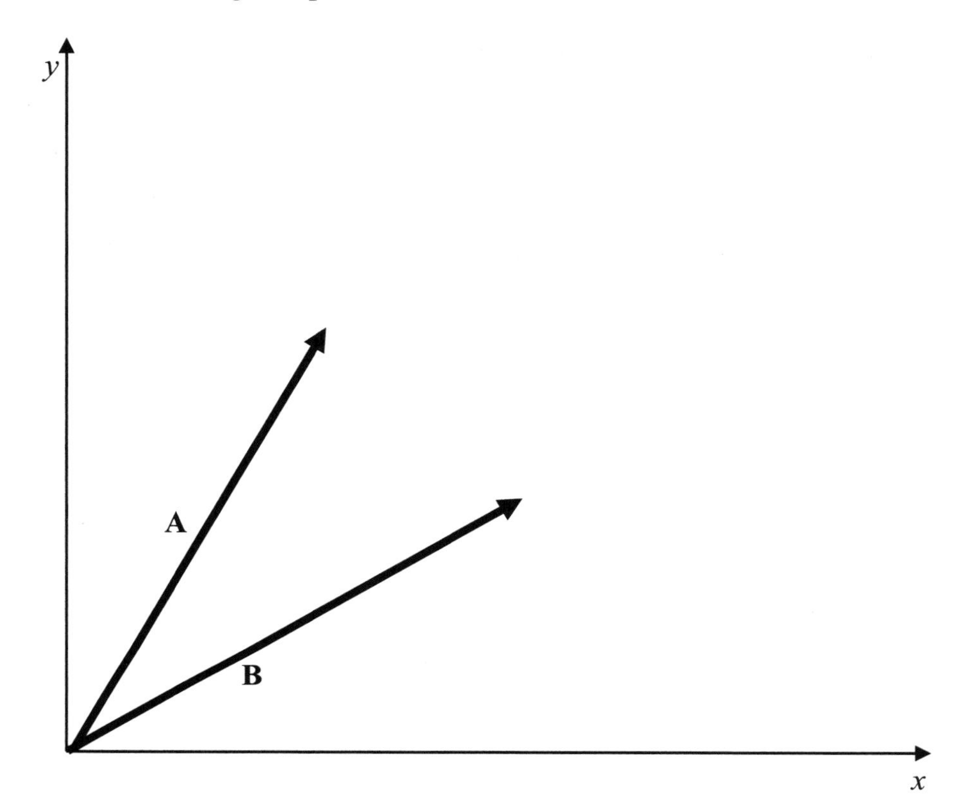

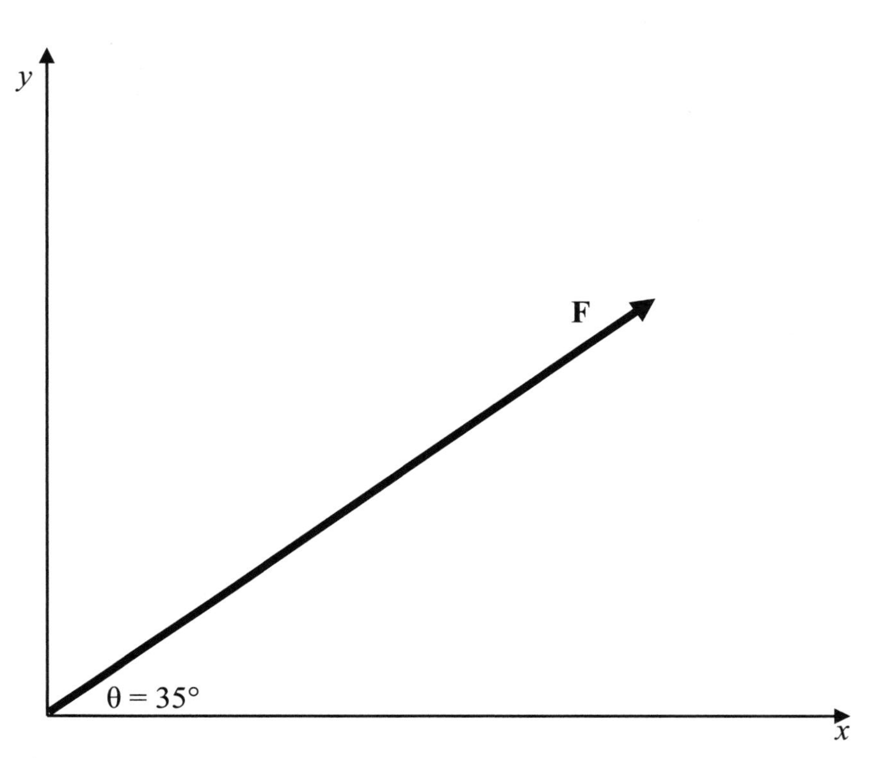

Name ________________________ Date ____________ Class ____________

7. A force vector **F** is shown in the picture above. The scale is 1 cm = 3 N.

   a. Use a ruler to find the length of the vector.

   b. Use the scale of 1 cm = 3 N to calculate the magnitude of the force.

   c. Calculate the x-component and y-component of the force **F** in Newtons.

8. A second coordinate system (labeled x' and y') is tilted 20° counterclockwise from the original coordinate system from the previous problem. Calculate the x'-component and y'-component in Newtons. (Hint: find the angle the force makes with respect to the new coordinate system first.)

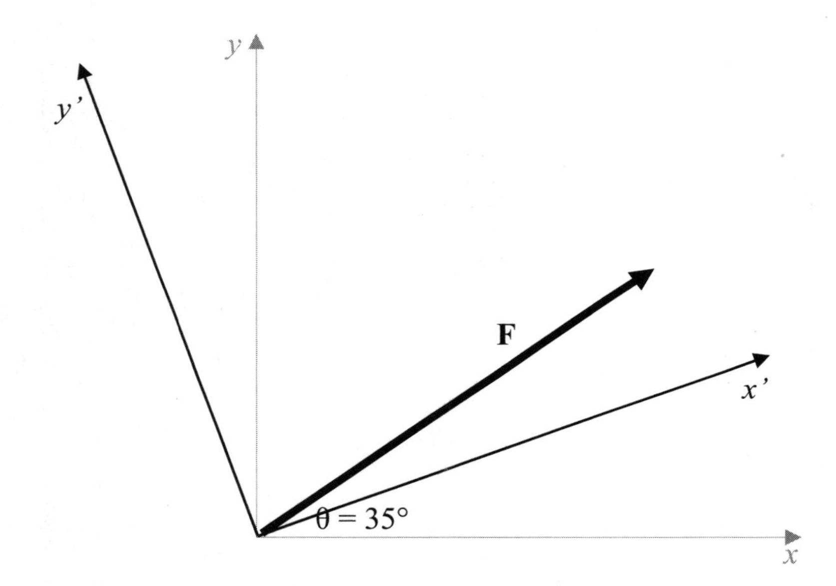

Name ______________________________ Date ____________ Class ____________

Motion in Two Dimensions

9. A strobe picture of a ball rolling at a constant velocity is shown below.

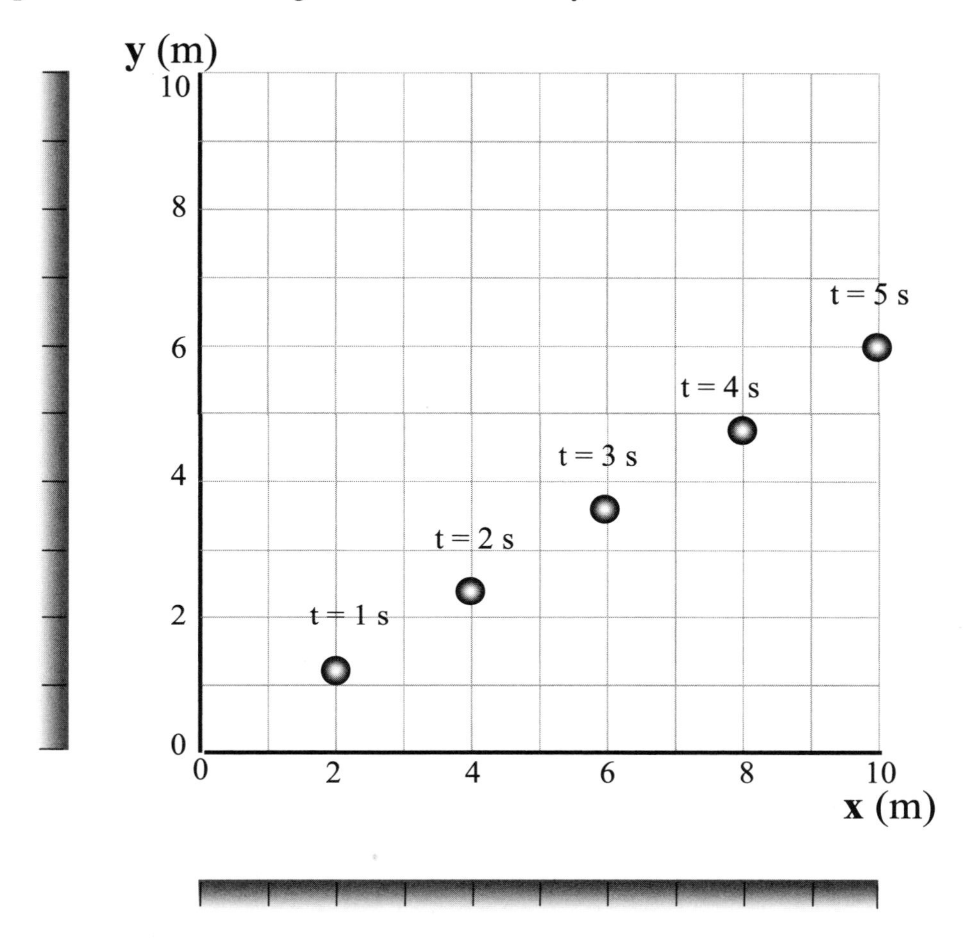

a. Draw the strobe pictures for motion in the x-direction and in the y-direction.

b. Is the x-component of the velocity, v_x, or the y-component, v_y, of the velocity larger? How can you determine this from the strobe picture?

c. Calculate v_x and v_y.

d. How can you find the magnitude of the velocity v from the components v_x and v_y?

e. Calculate v.

f. Under what conditions can v be smaller than v_x or v_y?

g. When can v be equal to v_x or v_y?

10. A strobe picture of a ball rolling at a constant velocity is shown below.

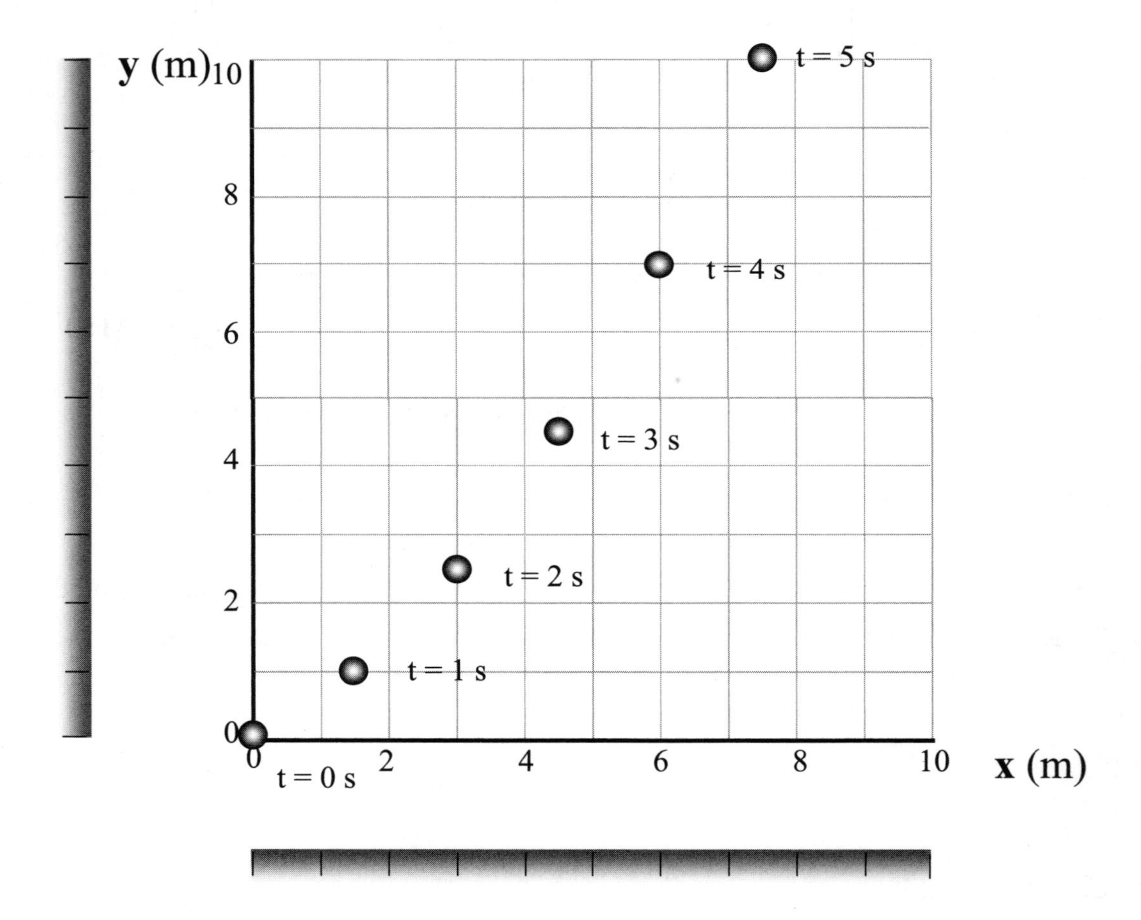

Name ______________________________ Date ____________ Class ____________

a. Describe the motion in the x-direction and y-direction (e.g. constant velocity, speeding up, etc.).

b. Draw the strobe pictures for motion in the x-direction and in the y-direction.

c. Plot v_x versus t and v_y versus t.

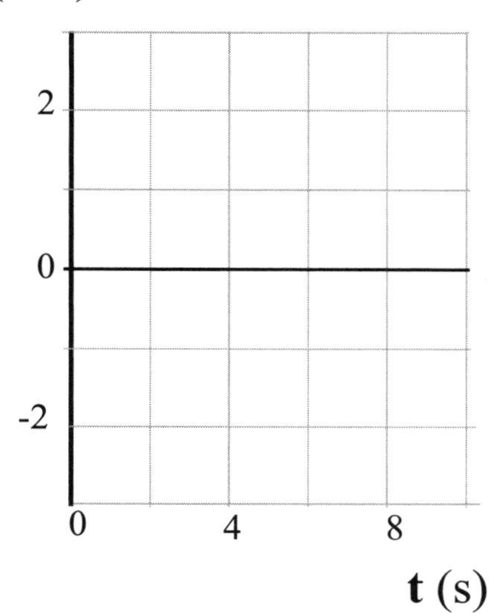

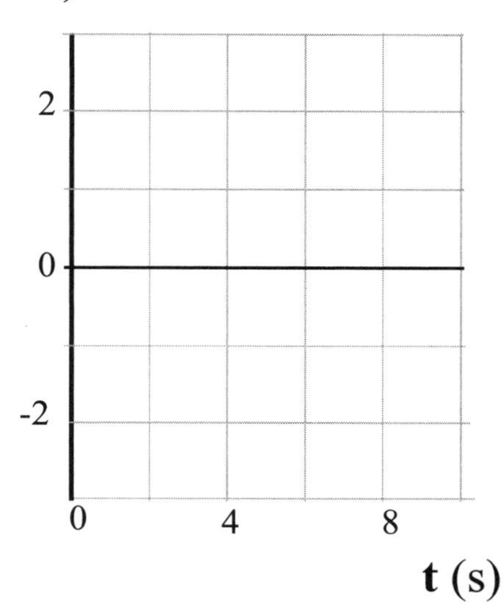

d. Calculate the constant acceleration or constant velocity along the two axes.

e. Sketch the path of the ball if the acceleration were the same in both directions on the following graph (i.e., $a_x = a_y$).

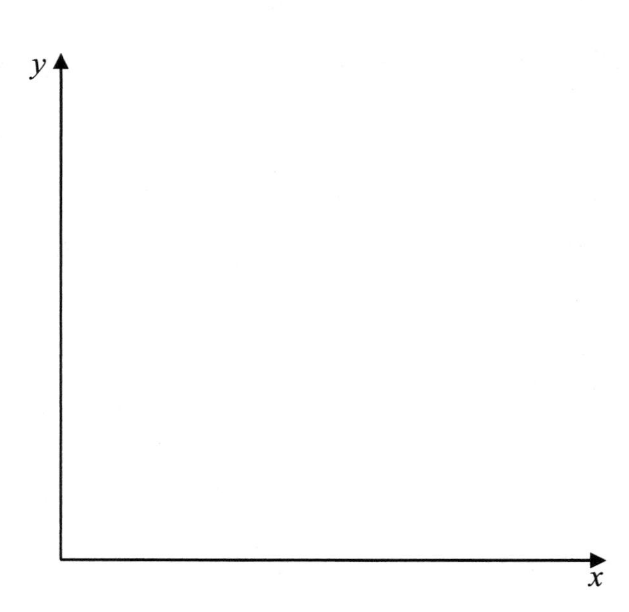

f. Sketch the path of the ball if $a_x > a_y$

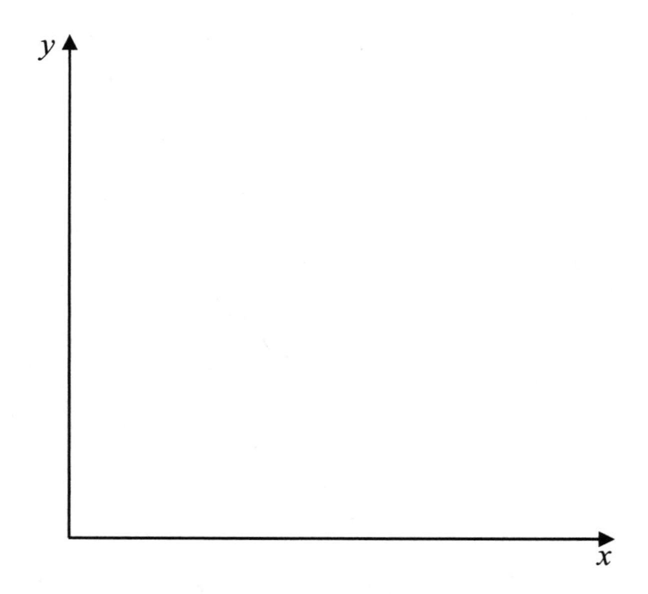

11. A cannon on level ground is angled up at 40°. If the cannonball has a muzzle velocity of 60 m/s, find how high the ball goes and how far from the cannon it lands.

    a. You can assume air friction is negligible. Why is this assumption valid?

    b. How fast is the cannonball moving horizontally?

Name ______________________________ Date ____________ Class ____________

c. How fast is the cannonball moving vertically when it exits the cannon?

d. In which direction is the cannonball moving with a constant velocity?

e. What is the acceleration in the other direction?

f. Plot v_x versus t and x versus t.

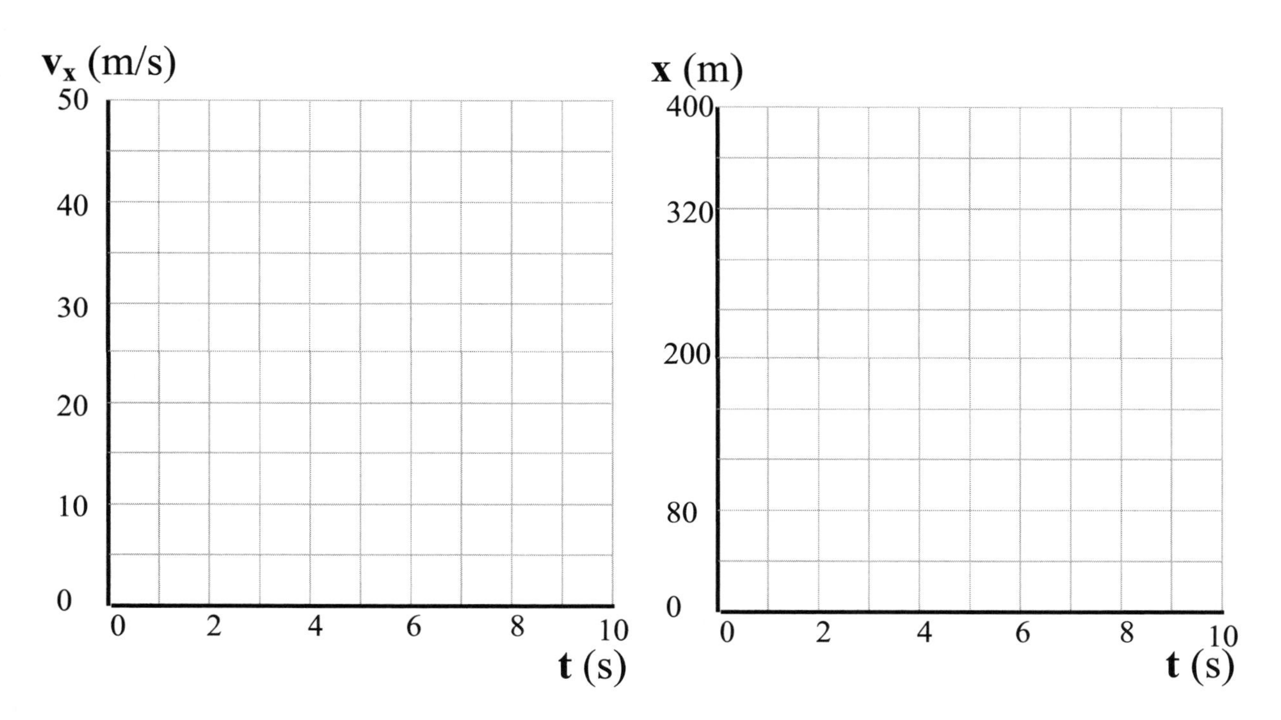

g. Can you determine the maximum height or total distance traveled using only these two graphs? Explain.

h. Which equation will you use to plot y versus t?

Name ______________________________ Date ____________ Class ____________

i. Plot v_y versus t and use the equation to plot y versus t.

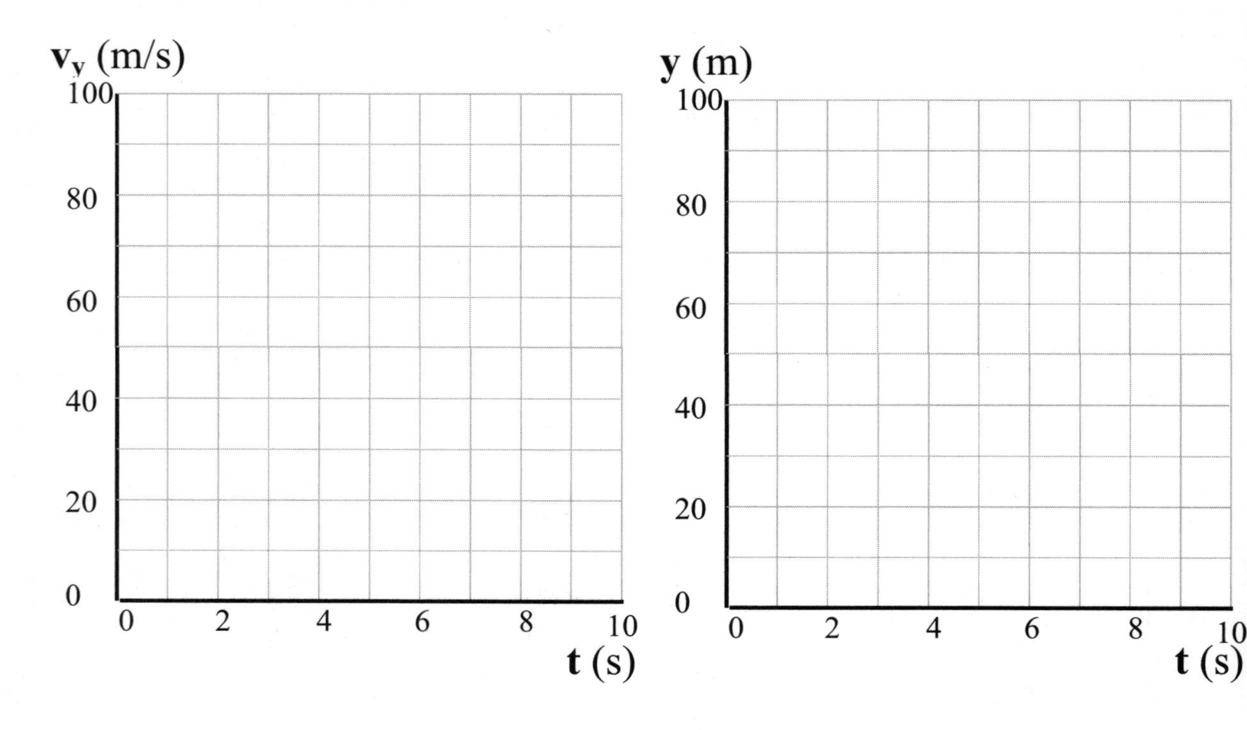

j. Can you determine the maximum height using only these two graphs? Explain.

k. From these two graphs, you can determine the time it takes for the ball to land, assuming the cannonball lands at the same height it started, but you cannot determine the total distance traveled. What other information do you need to find the total distance traveled?

l. The vertical velocity v_y is zero at the peak of the trajectory. Does zero velocity imply the acceleration is also zero? Explain.

Name ________________________________ Date ____________ Class ____________

m. Use your x versus t graphs to sketch the strobe pictures in the horizontal and vertical direction. Use these strobe pictures to draw the path the cannonball takes. (Note: the horizontal and vertical scales are different.)

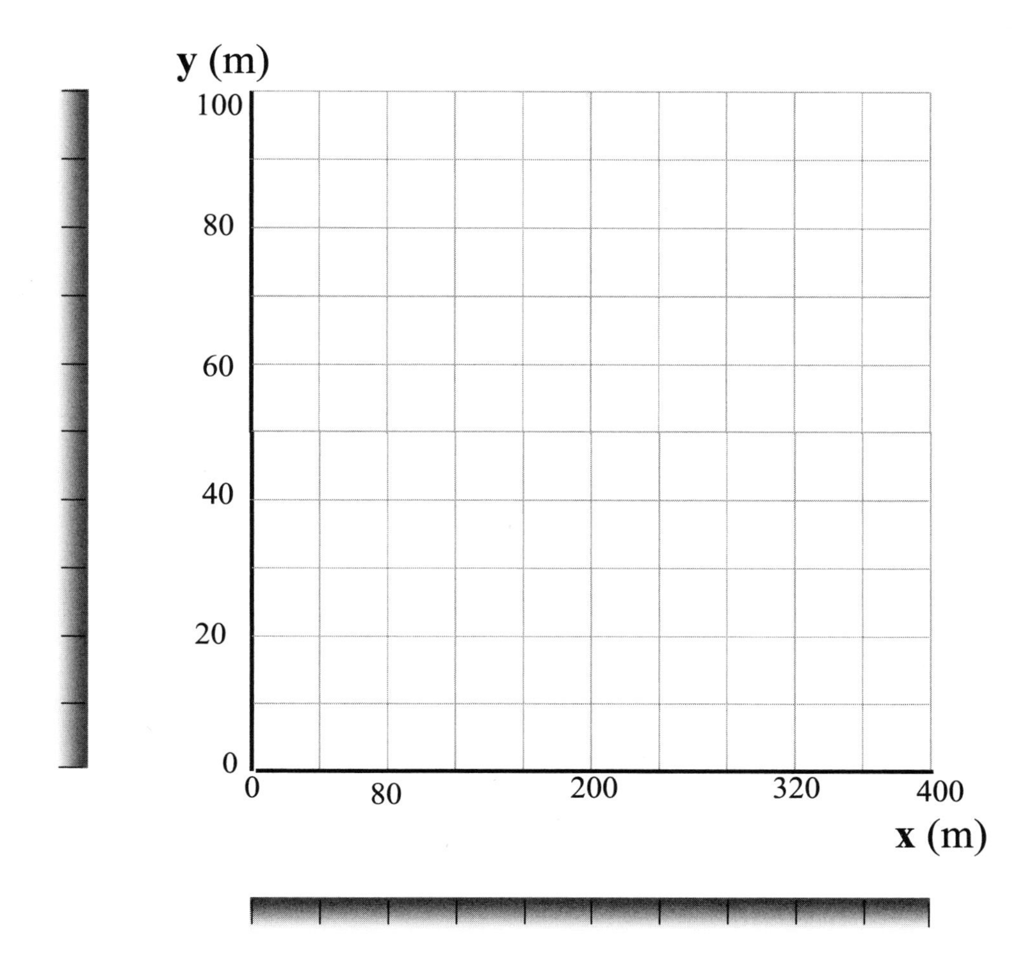

n. What is the maximum height of the ball?

o. How far does the ball travel horizontally?

Name ______________________________ Date ____________ Class ____________

p. Imagine your friend is working on this problem and needs your help. Write a brief description of the steps you'd need to solve the problem.

Name ______________________ Date ____________ Class ____________

Chapter 4

Interactions and Newton's Laws of Motion

1. Ann and Bob are in a car traveling at a constant velocity of 65 mph. Bob says, "Since our velocity is constant, there cannot be anything pushing or pulling on the car." Ann responds, "There are a lot of things pushing and pulling on the car, but as long as they all balance, the car moves with a constant velocity."

   a. What is wrong with Bob's comment?

   b. Restate Ann's response in terms of the sum of the forces on the car.

2. Two students are arguing about a ball that is moving at a constant velocity. Sarah says, "There must be a force pushing the ball forward, otherwise it would come to a stop." Tina says, "All of the forces exerted on the ball must balance, since its velocity is constant."

   a. Does a moving object require a force to continue moving? Explain why or why not.

   b. What is wrong with the remarks made by Sarah?

   c. Explain what is wrong with the following statement: "Since the ball moves at a constant velocity, there can be nothing pushing or pulling it."

d. What conclusion about the sum of the forces on the ball can you draw from the fact that the ball moves with a constant velocity?

3. Two students are having a disagreement about baseball. Ranjani says, "The baseball will continue to move until it hits something." Samantha says, "The baseball continues to move because of the force of the baseball bat." How would you settle this argument?

4. As you watch a bowling ball roll down the lane, you notice it slows down slightly. What conclusion about the sum of the forces on the bowling ball can you draw from the fact that the speed of the ball changes?

5. You gently toss a baseball into the air and take a horizontal swing at it with a bat and repeat this with a ping pong ball.

   a. When the bat strikes each ball, the baseball exerts a larger force on the bat than the ping pong ball does. Is this due to the weight difference of the two balls or the difference in mass? Explain.

   b. Does inertia exert a force on the ball? Explain.

   c. What horizontal forces do the balls experience while colliding with the bat?

Name ______________________________ **Date** ____________ **Class** ____________

6. A tomato is tossed into the air. As it descends, you swing a knife horizontally and cut the tomato in half. Is there a horizontal force holding the tomato in place so the knife can cut through it? Explain your reasons for choosing your answer.

7. Two of your friends are talking about an incident when their car ran out of gas. George says it was difficult for the two of them to push the car because it weighed so much. Kim contends that it was hard to push because it was so massive. Which of your friends is wrong? How would you convince your friend of the correct answer?

8. Two friends, Butch and Cal, are standing back-to-back when Butch pushes off a nearby tree, causing both friends to fall down. Butch says, "It's your fault we fell over because you were pushing on me with a force smaller than me pushing on you." His friend Cal says, "I was pushing on you just as hard as you pushed me, but you pushed off the tree and I had nothing to oppose your push."

   a. In modern English, what does Newton's third law state?

   b. Do Butch's comments make sense in light of Newton's third law? Explain your reasoning.

Name ______________________________ Date ____________ Class ____________

9. Michelle and Nick are debating the forces between a train engine and the first train car. Nick contends, "If the train is speeding up, the force exerted by the engine on the first car has to be larger than the force of the first car on the engine, otherwise the forces would cancel and the train wouldn't budge." Michelle says, "As long as the force the first car exerts on the engine is smaller than the force between the engine and the track, the train will accelerate."

   a. Why doesn't the force of the engine on the first car cancel the force of the car on the engine?

   b. Explain to Nick why third law interaction pairs do not cancel each other.

10. A large stone is dropped and lands on top of a brick lying on a table.

   a. Is the force the falling stone exerts on the brick ever different in magnitude than the force the brick exerts back on the stone? Explain your reasons for choosing your answer.

   b. When the stone hits the brick on the table, both objects break through the table. Does the falling stone directly exert a force on the table?

   c. Is the force exerted by the table on the brick ever greater than or less than the force the brick exerts on the table (a) before the stone hits the brick, (b) while the brick is breaking through the table, or (c) after the brick has broken through the table? Explain.

Name ______________________________ Date ____________ Class ____________

11. Two students are discussing the forces on the bottom book in the diagram. Danni says, "My hand isn't touching the bottom book, so there can't be a contact force between them." Eric says, "When you press down hard, I can see the bottom book compress a little, so I know your hand is exerting a force on the bottom book."

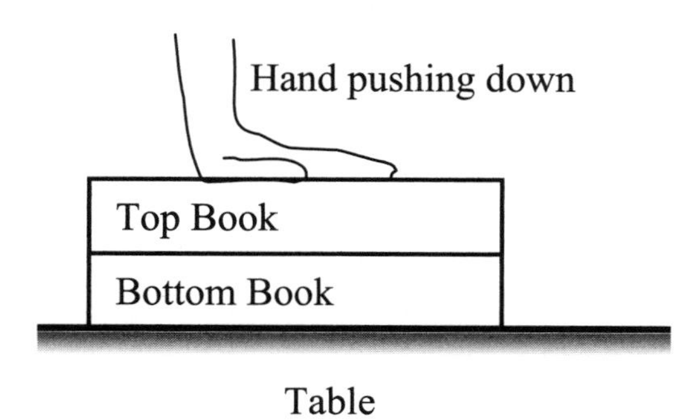

a. Since the hand is not in contact with the book, is the force from the hand on the bottom book an action-at-a-distance force?

b. Does the top book push down on the bottom book with a force equal to its weight or with a larger force?

c. If two more books were stacked on top of the top book, would the force increase on the bottom book?

d. Would the two books exert a force directly on the bottom book?

e. Explain to Eric why Danni was correct.

Name ______________________ **Date** ____________ **Class** ____________

12. The two boxes shown in the diagram are stationary and are in contact with one another.

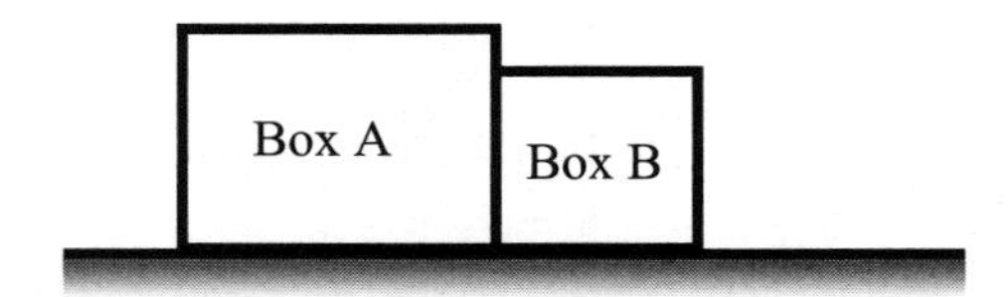

a. If you remove box A, does it have any effect on box B?

b. Is there a force exerted by one box on the other?

13. The two boxes, stacked one on top of the other, are stationary.

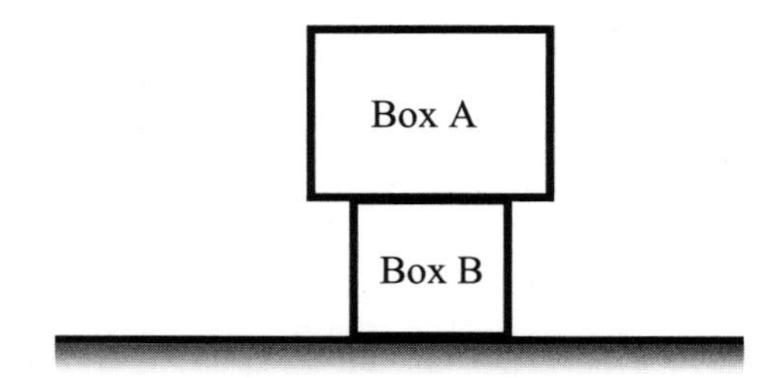

a. If you remove box B, does it have any effect on box A?

b. Is there a force exerted by one block on the other?

c. Is the magnitude of the force exerted by A on B smaller than, the same as, or larger than the force B exerts on A? Explain your answer.

Name ______________________________ Date ____________ Class ____________

14. Box B is leaning against box A. Neither box is moving.

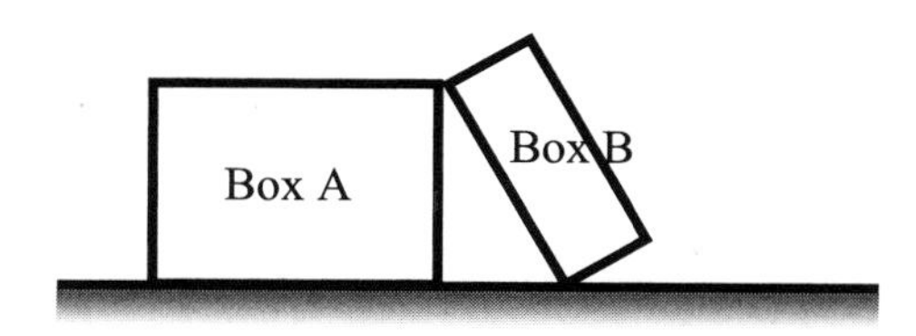

a. If you remove box B, does it have any effect on box A?

b. Is there a force exerted by one box on the other?

c. Is the magnitude of the force exerted by A on B smaller than, the same as, or larger than the force B exerts on A? Explain your answer.

15. Based on your answer to the previous questions, what is a good test for whether two objects exert a force on one another?

16. Sketch the *v* versus *t* graph and the force *F* versus *t* graph based on the *x* versus *t* graph.

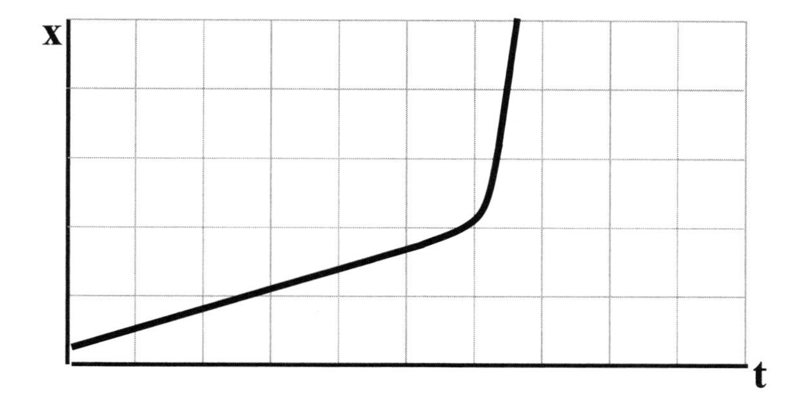

Name ______________________________ Date ____________ Class ____________

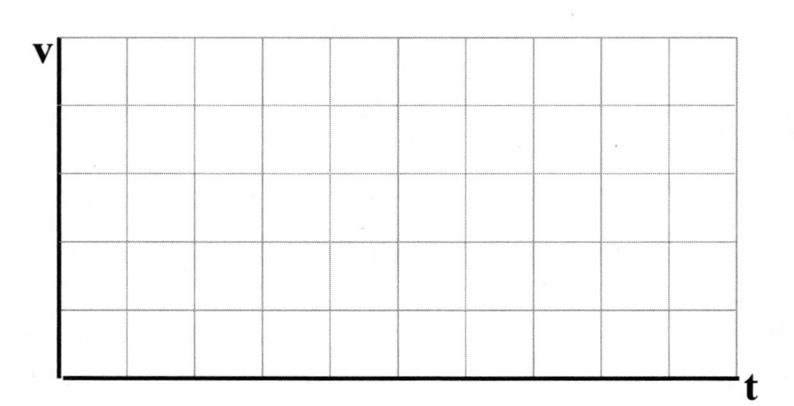

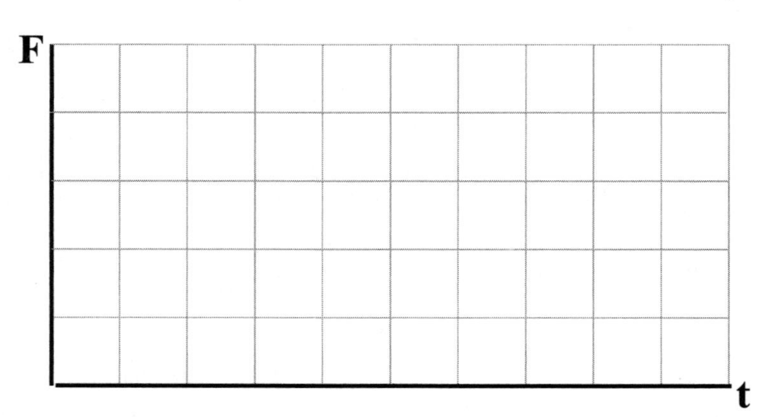

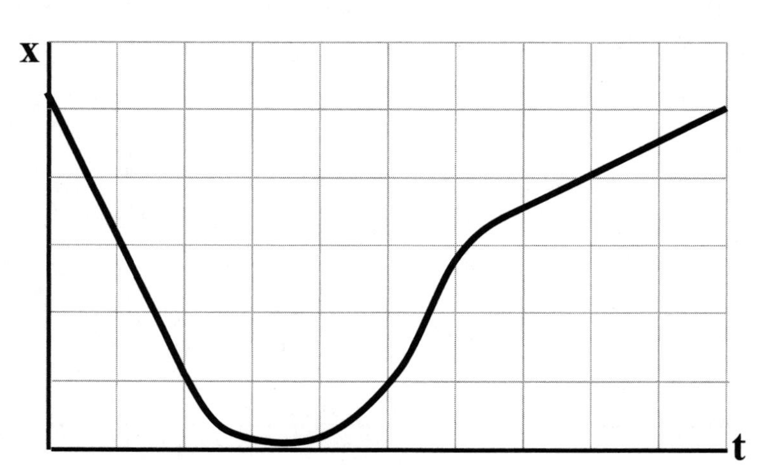

Name ______________________________ Date ____________ Class ____________

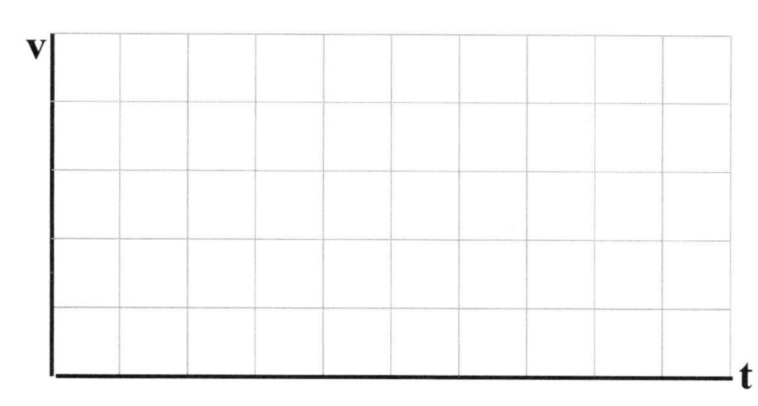

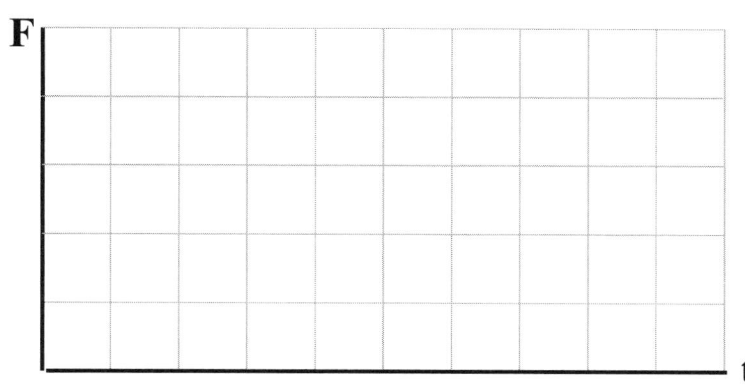

17. Sketch the force *F* versus *t* based on the *v* versus *t* graph.

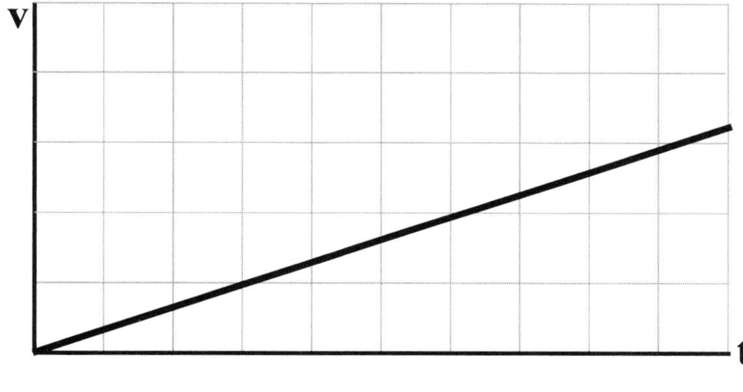

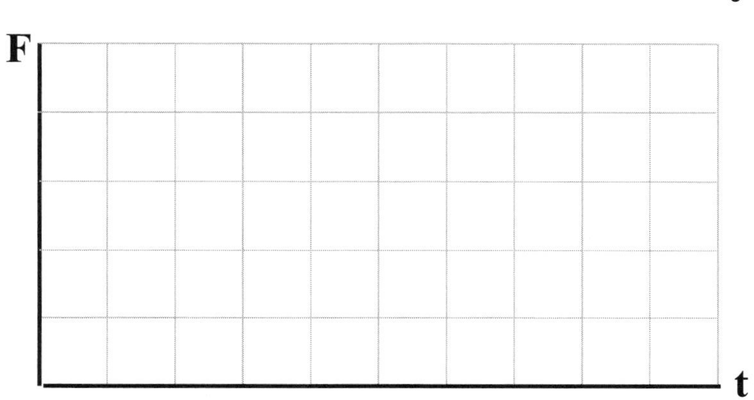

Name ______________________________ Date ____________ Class ____________

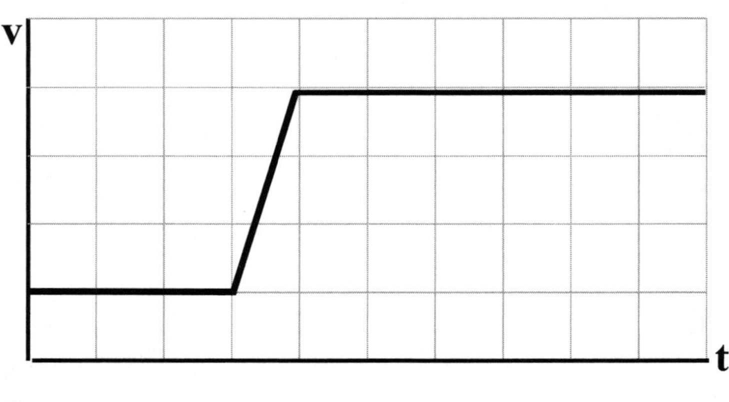

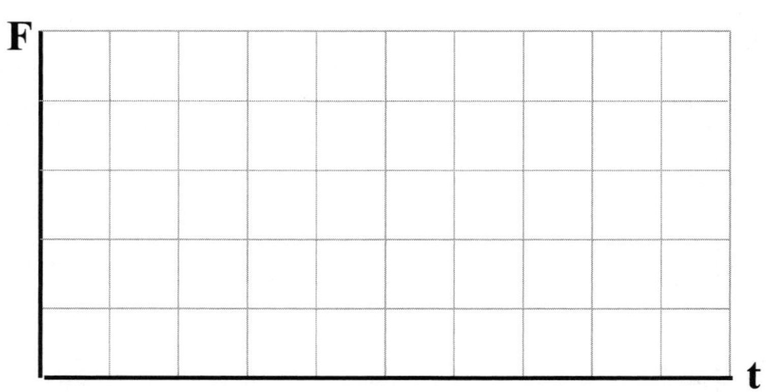

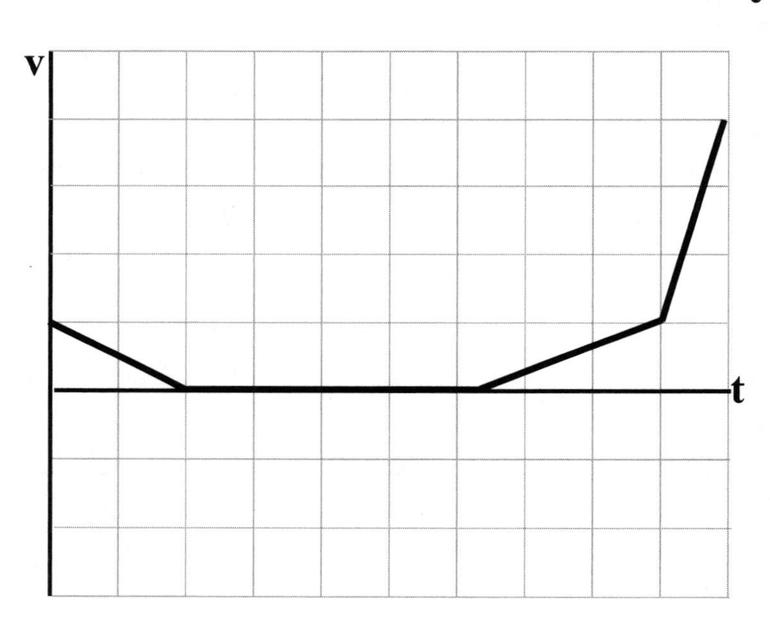

Name ______________________ Date __________ Class __________

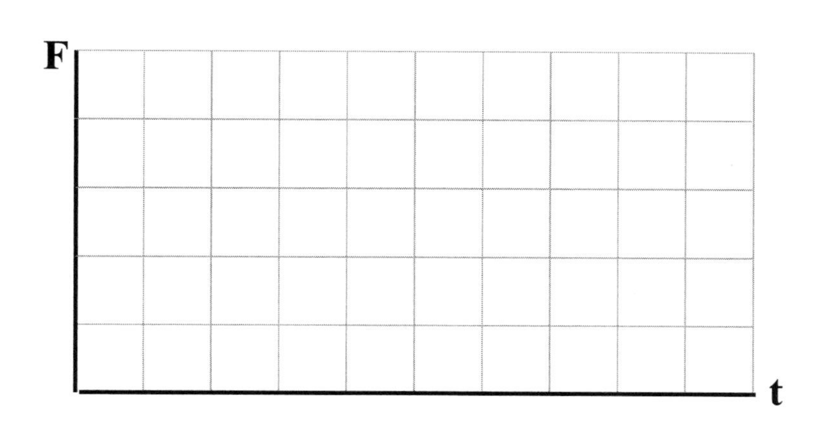

18. Circle the regions where the ball is experiencing a net force. If the ball does not appear to experience any forces, write "Vector sum of forces is zero." Explain how you can tell the ball is experiencing a net force in each case.

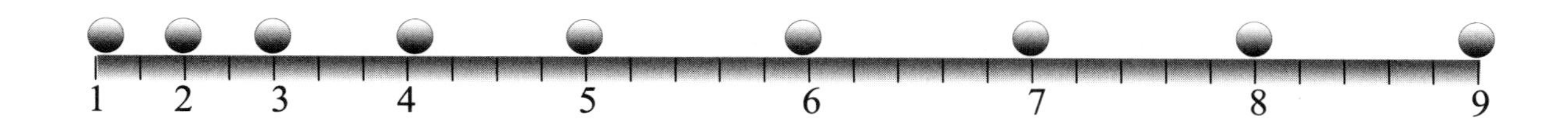

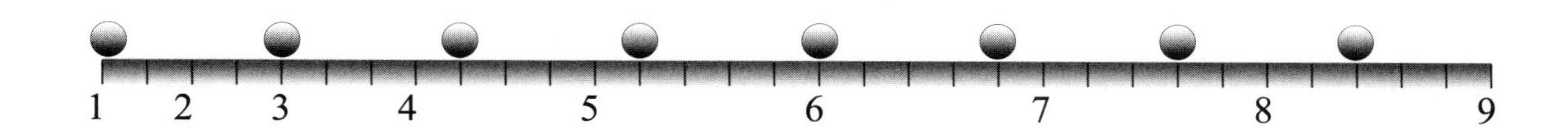

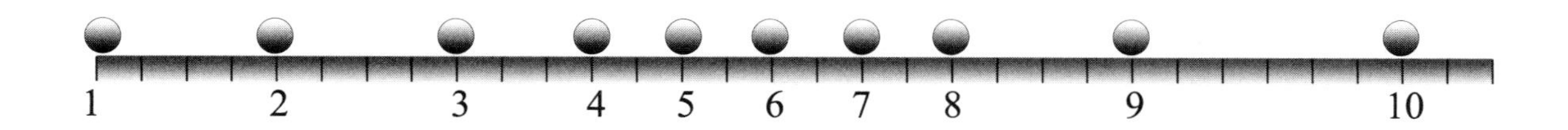

Name ______________________________ **Date** ____________ **Class** ____________

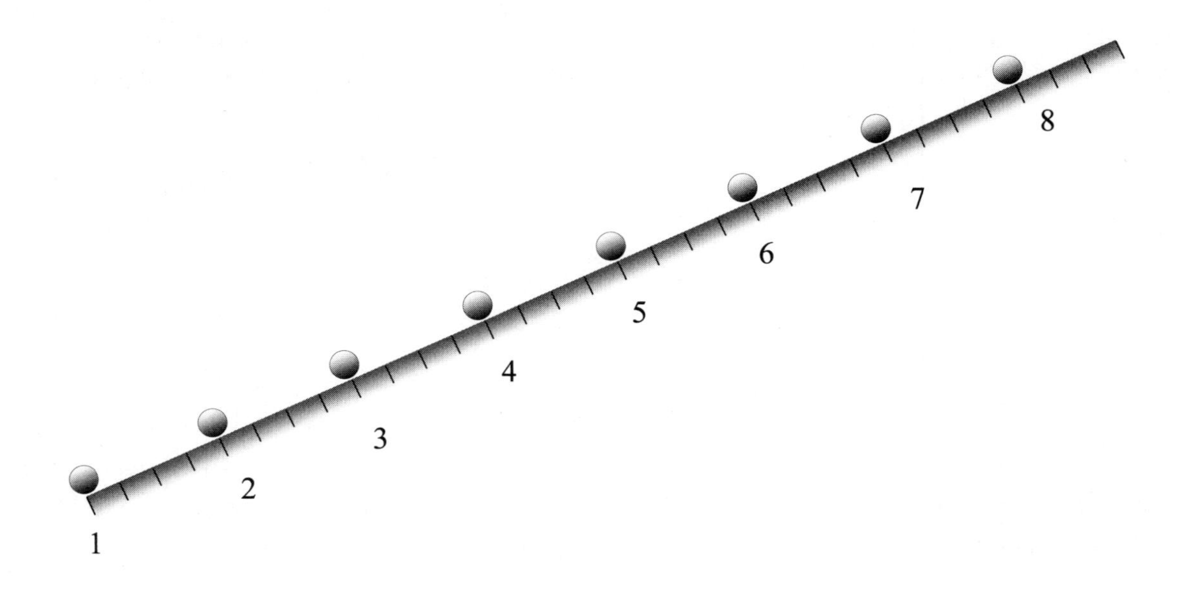

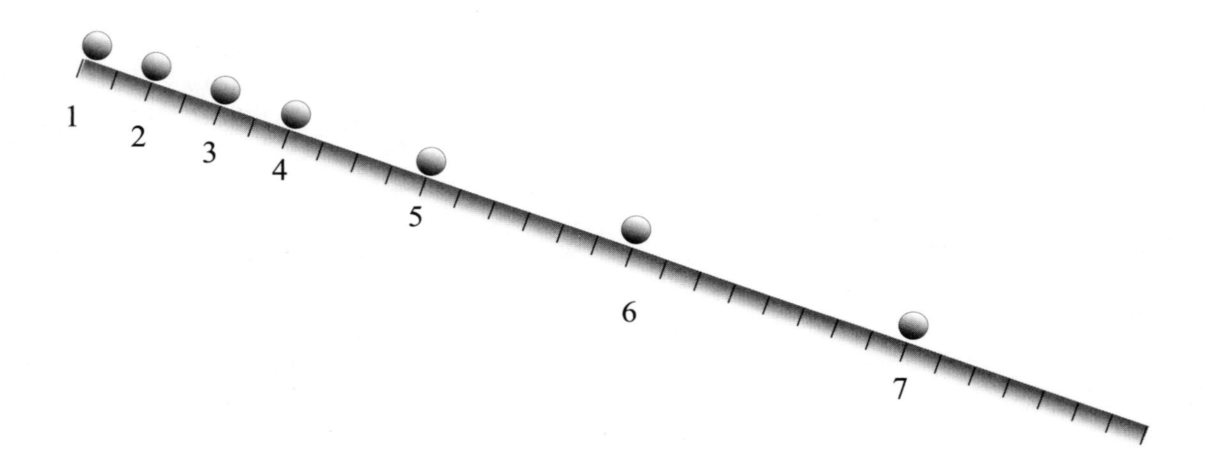

Name ______________________________ Date ____________ Class ____________

19. Explain at which points each vehicle experiences a net force in the following scenarios. Specify when the net force is first applied and when the net force becomes zero again. Explain your reasoning.

   a. A car is initially idling at a stop light. The light turns green and the car takes off until it reaches a cruising speed of 45 mph. Up ahead, the driver sees slow moving traffic and she slows down to 20 mph.

   b. A car going 35 mph rounds a bend at a constant speed and then accelerates to 65 mph. After driving at 65 mph for a few minutes, the car speeds up to pass a slow moving truck. After passing the truck, the car resumes traveling at 65 mph.

   c. A small airplane taxis from its hangar and turns on to the runway. It accelerates and lifts off. The plane climbs at a constant speed to its cruising altitude. After a bit, it turns around to head back to the airport. After a few more minutes, it starts a slow descent until its wheels touch down. The pilot applies the brakes until the plane rolls to a stop.

20. A 0.75 kg ball is launched horizontally from a cannon. The average force exerted on the ball by the cannon is 24 N.

   a. How much force would be necessary to launch the ball with the same horizontal velocity as on the moon, where the weight of the ball is roughly $1/8^{th}$ of its weight on Earth?

   b. How much force would be needed to launch the ball in deep space, where its weight is zero?

   c. Which determines the difficulty in accelerating the ball: weight or mass? Explain.

Name ______________________________ Date ____________ Class ____________

21. In the game of curling, a 19 kg stone is tossed along a 45.7m (150 ft) ice track. The object is to toss the stone so it comes to a stop on a target at the far end of the track.

    a. What causes the stone to come to a stop?

    b. As the stone moves down the ice, two players run in front of the stone scrubbing at the ice with brooms. Why does scrubbing the ice allow the stone to travel farther?

    c. If the stone experienced no friction as it slid across the ice, how far would the stone travel before it began to slow down?

    d. Can a moving object come to a stop if there are no forces exerted on it?

22. A book with a mass of 3 kg is suspended by a rope.

    a. Why doesn't the book fall? What is the force called that keeps it from falling?

    b. What is the net force on the book?

    c. What force exerted by the rope on the book is necessary to hold the book in place?

Name ______________________________ Date ____________ Class ____________

23. You start pulling up on the rope, applying a constant force of 40 N. Answer the following questions to find the acceleration of the book.

   a. Draw a rough sketch of the book with a rope attached. Draw and label the forces acting on the book.

   b. If the problem implied the book were stationary, you could assume the net forces were zero. However, nothing in the problem leads us to believe the book is stationary. What is the net force on the book?

   c. Now that you know the net force on the book, which of Newton's laws can you use to find out the acceleration of the book?

   d. Calculate the acceleration of the book.

   e. Compare your answer to the acceleration if the book were falling freely.

f. What would the gravitational force on the book be if it were falling freely?

g. What would the book's acceleration be if it w falling?

h. Would you expect your answer to be very large, very small, or somewhat similar to the acceleration of a falling book?

i. Does your answer seem reasonable?

Name ______________________ Date ____________ Class ____________

Chapter 5

Problem Solving Using Newton's Laws

1. A book is sitting on a table at rest.

   a. Are there any forces in the horizontal direction exerted on the book?

   b. What two forces in the vertical direction does the book experience?

   c. Which force is a contact force? Which is an action-at-a-distance force?

   d. Draw all the forces exerted on the book resting on a table. Be sure to label each force (e.g. $F_{\text{on book by table}}$).

   e. Is one force larger than the other? Explain.

f. The book is stationary on the table. What conclusion about the sum of forces on the book can you draw?

g. What objects does the book exert a force on? Match each force exerted by the book with its reaction pair force, exerted on the book.

2. A block is stationary on a ramp.

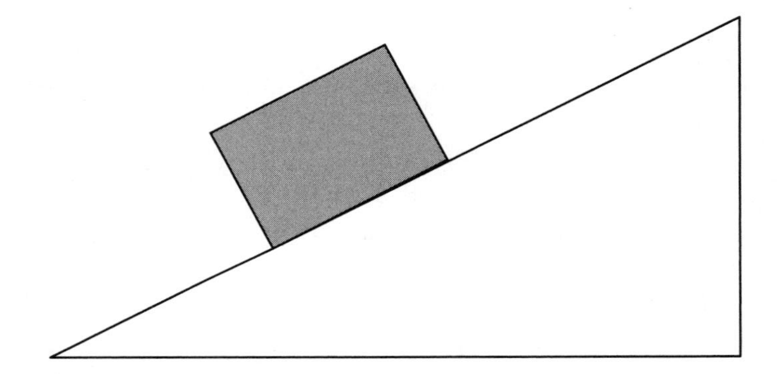

a. How can you deduce the net force experienced by the block?

b. Which direction is the ramp pushing on the block?

c. Which direction is the frictional force on the block?

d. Draw all the forces exerted on a stationary block sitting on a ramp. Clearly label each force.

e. Rank the forces exerted on the block from smallest to largest.

3. A cart is rolling with a constant velocity along a surface with negligible friction.

   a. What is the net force in the horizontal direction? Explain how you got your answer.

   b. Since friction is negligible, you do not have to consider it. Based on your answer to the previous question, are there other horizontal forces exerted on the cart?

   c. What are the forces in the vertical direction? What type of force is each?

Name ______________________________ Date ____________ Class ____________

d. Draw the forces on the diagram below and label each vector (e.g. $F_{\text{on cart by ground}}$).

4. A cart is pushed with a constant force along a surface with negligible friction.

   a. Since friction is negligible, the only other force is from the hand pushing. Based on this fact, can you tell whether the cart is stopped, moving at a constant velocity, or accelerating? Explain.

   b. Should you include a force = m × a in the forward direction? Explain why or why not.

   c. What are the forces in the vertical direction? What type of force is each?

Name ______________________ Date ____________ Class ____________

d. Draw the forces on the diagram below and label each vector (e.g. $F_{\text{on cart by ground}}$).

5. A block, initially at rest, starts to slide down an incline ramp.

   a. List the forces experienced by the cart and what type of force they are.

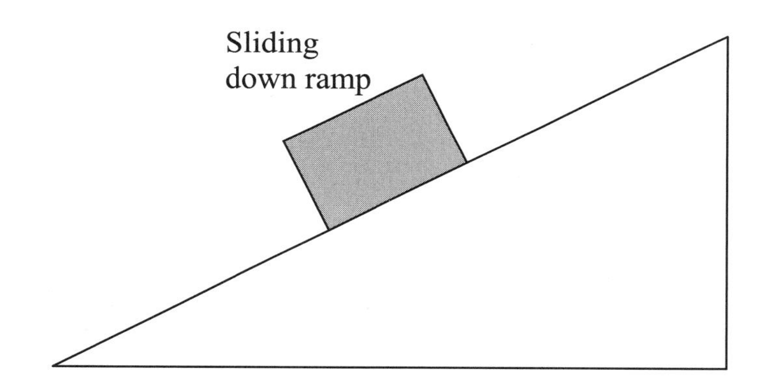

   b. Draw all of the forces exerted on the block as it accelerates down the ramp. Label each force as in previous exercises.

   c. Rank the forces exerted on the block from smallest to largest.

d. What objects does the block exert a force on? Match each force exerted by the block with its reaction pair that is exerted on the book.

Free Body Diagrams

6. Correct the diagram below and draw the correct free body diagram.

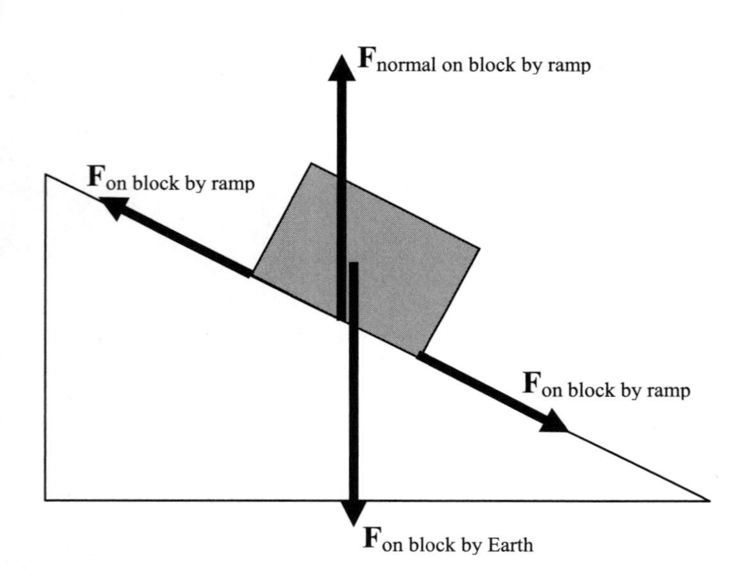

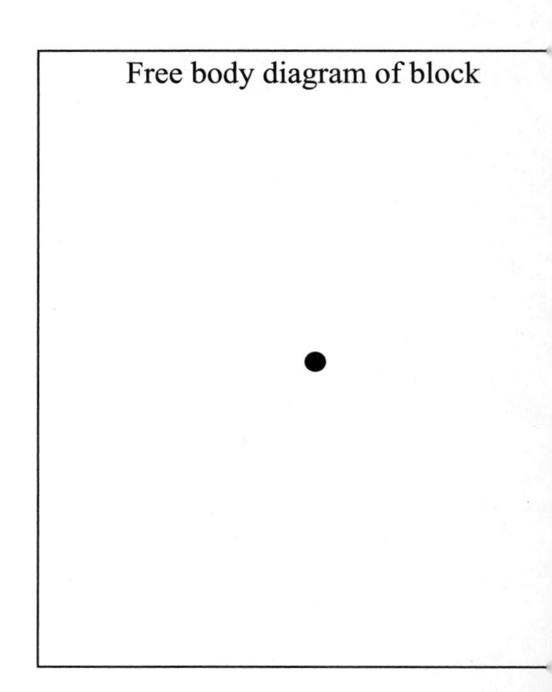

Name ______________________________ Date ____________ Class ____________

7. Correct the free body diagram for the bottom book.

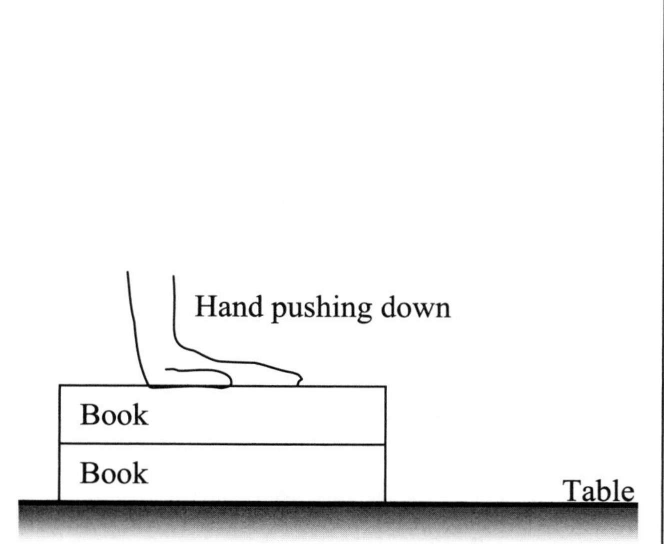

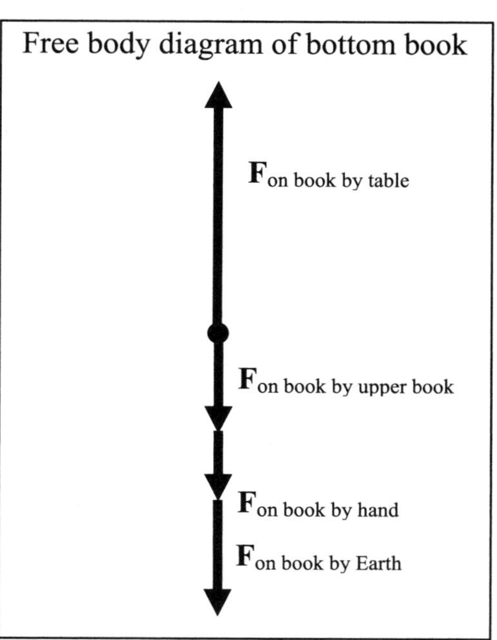

Name ______________________ Date __________ Class __________

8. A hand is pushing on the top book but both books remain stationary. Correct the free body diagram for both books.

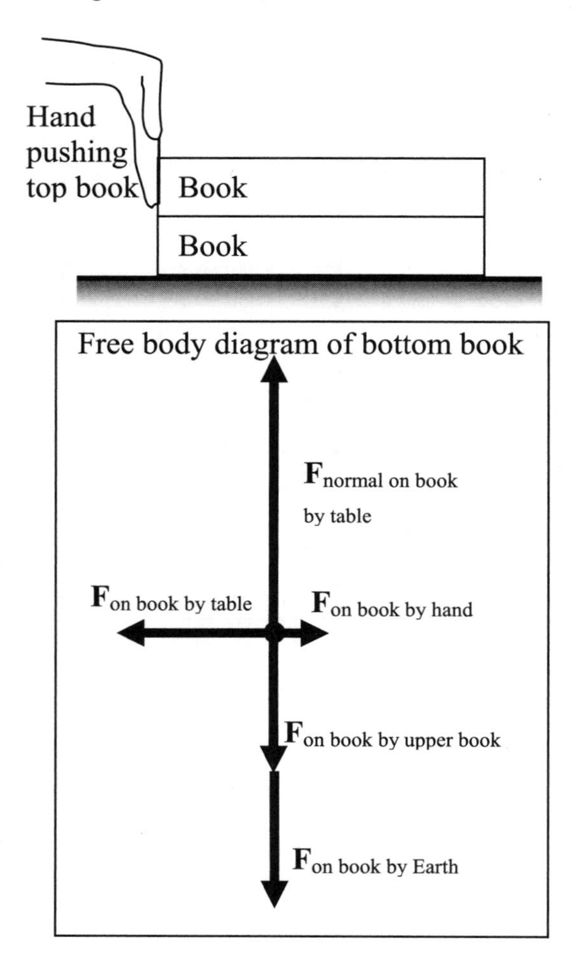

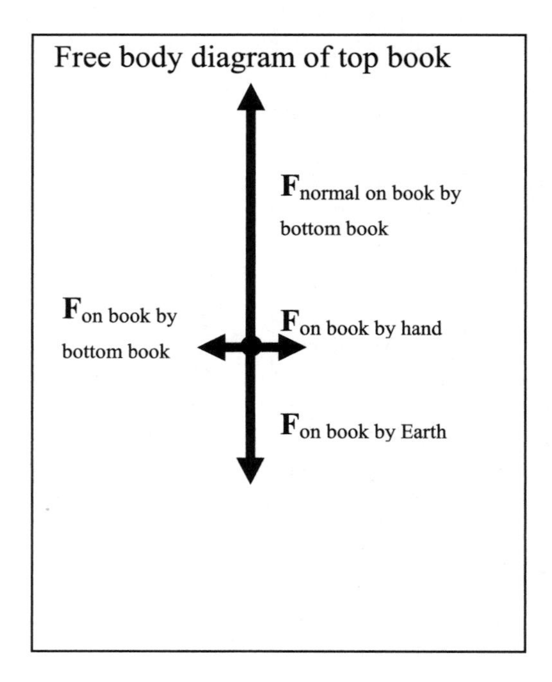

9. Draw a free body diagram for a book resting on a table.

Book

Table

Free body diagram for book

Name ______________________ Date __________ Class __________

10. Draw a free body diagram for a block sitting on a ramp.

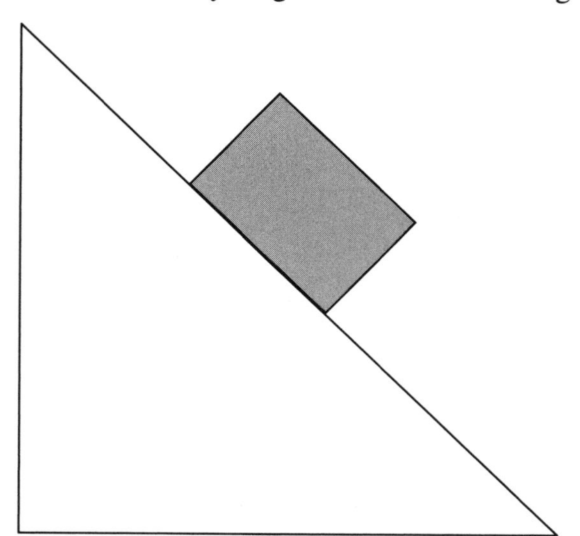

Free body diagram of block

11. Draw a free body diagram for a magnet stuck to the door of a refrigerator. Be sure to label each force.

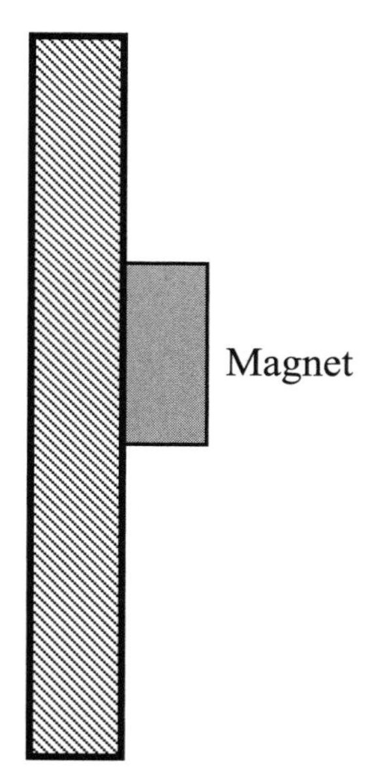

Free body diagram of magnet

Name ______________________________ Date ____________ Class ____________

12. You place your latest physics quiz on the fridge for display. Draw a free body diagram for the piece of paper and magnet.

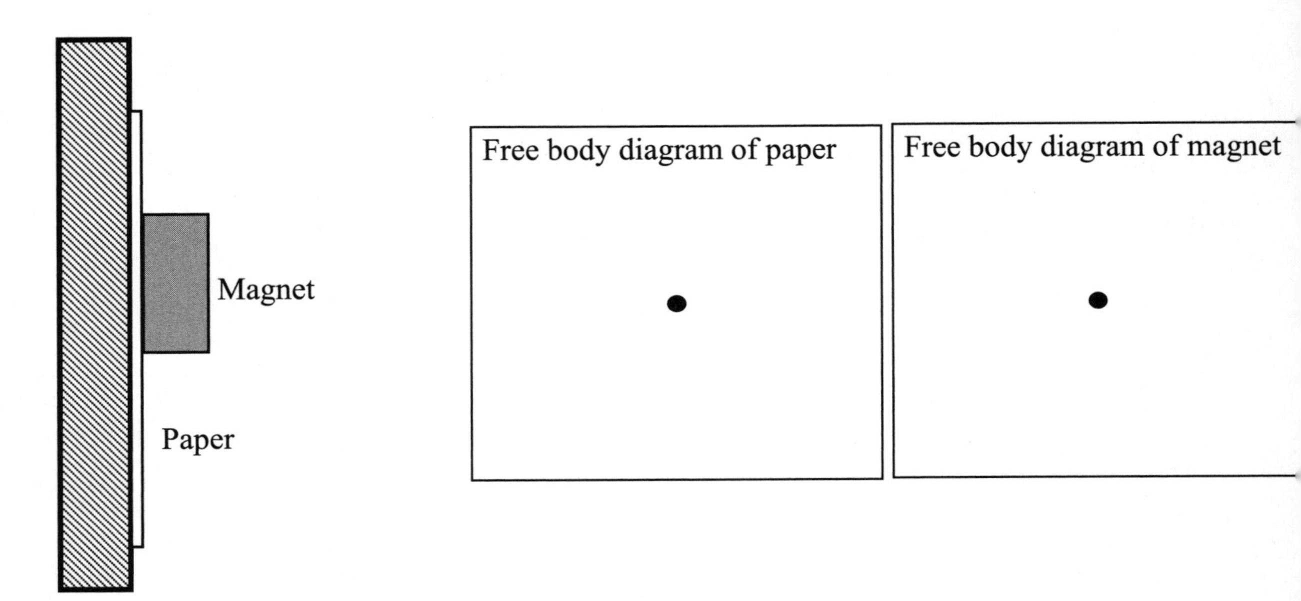

13. If you slowly pull on the paper, the magnet moves with the paper, but if you yank it out fast the magnet stays in place. Explain why this happens.

Third Law Pairs of Forces

14. Two blocks connected by a rope are being pulled and are accelerating toward the right. Draw a free body diagram for each block and the rope. Label the forces exerted on the two blocks and the rope (assume the mass of the rope is so small you can ignore it). Label all of the third law interaction pairs.

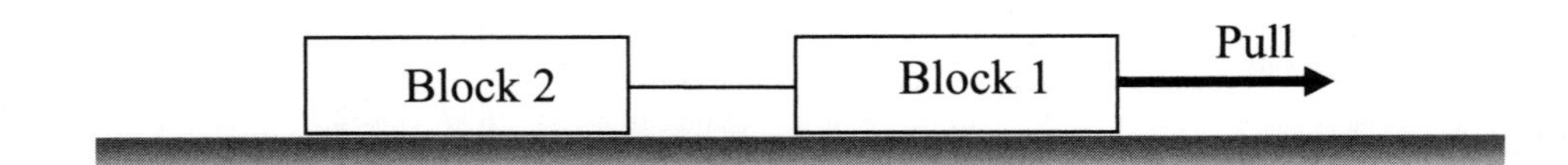

Name ______________________________ Date ___________ Class ___________

15. If the blocks are still being pulled but now are moving at a constant velocity, how would the free body diagrams change?

16. Redraw the free body diagrams for the sheet of paper and the magnet. Specify the third law interaction pair for each force.

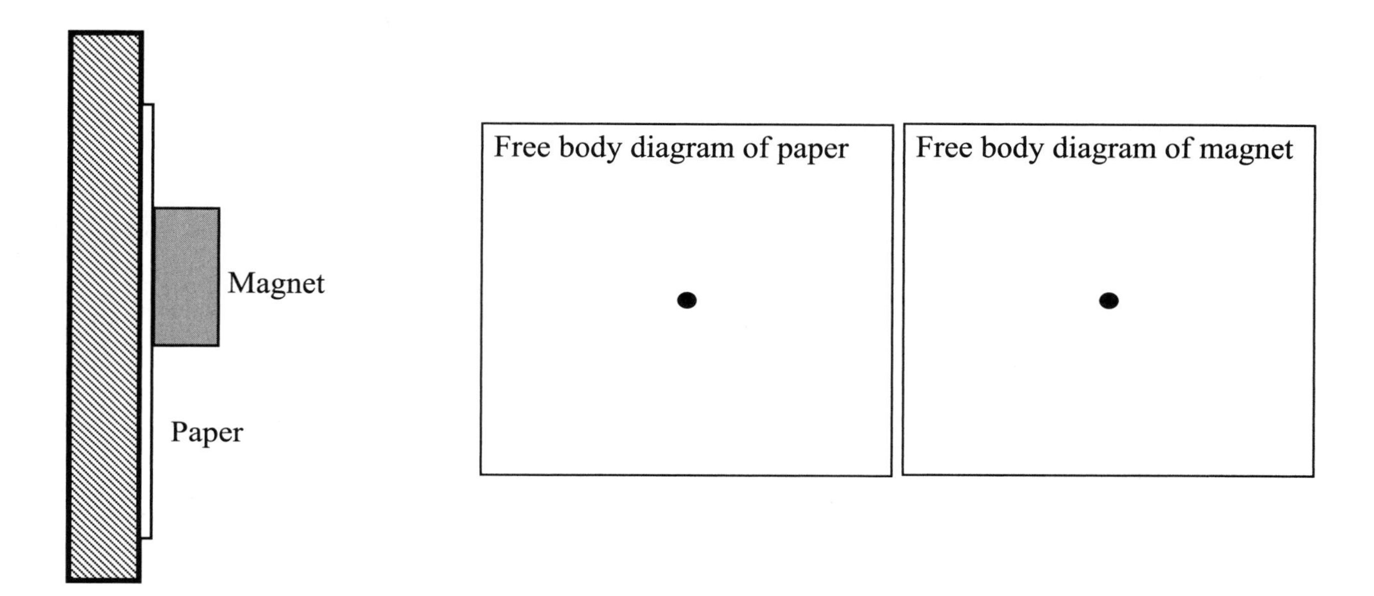

Tension

17. In terms of the mass of the block m and the acceleration due to gravity g, what does the force sensor read?

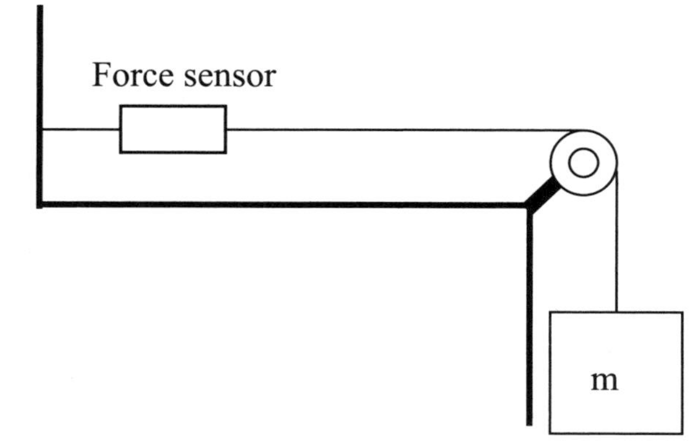

Name ______________________________ Date ____________ Class ____________

18. If the rope is cut at the point indicated and a sensor is stuck in there, what will the sensor read now? (The sensor is very light so you can ignore its weight.)

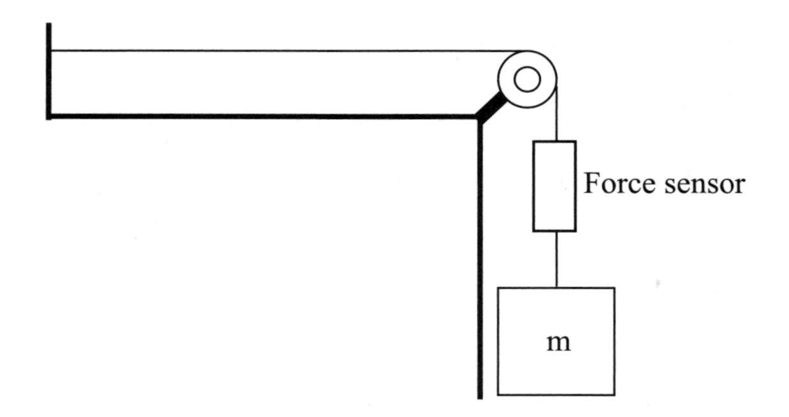

19. A pair of identical blocks is connected by a string that is run over a single pulley. Both blocks are stationary and have mass m. What does the force sensor read?

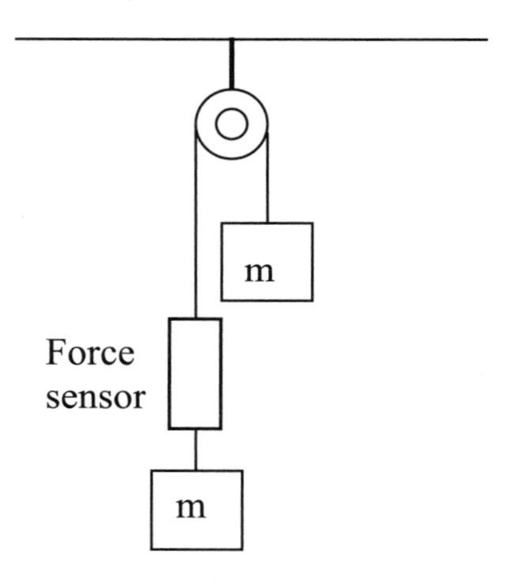

20. A single mass is hanging from a hook in the ceiling. What does the force sensor read?

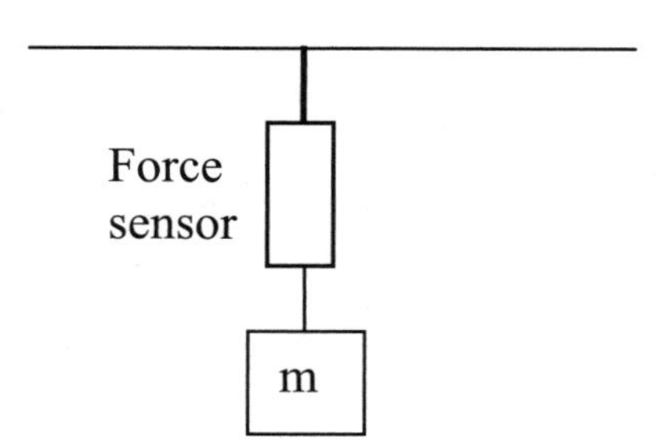

Name ______________________________ Date ____________ Class ____________

Torque

21. You are trying to loosen a nut with a long-handled wrench.

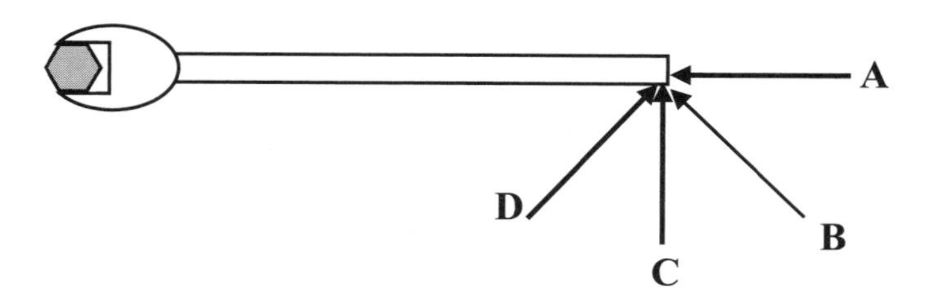

a. Why is pushing in the direction of **A** a waste of time?

b. Which direction would you push if you wanted to push as little as possible to loosen the nut? Explain why the other directions would require more effort.

Name ______________________ Date ____________ Class ____________

22. Two friends, Andy and Bill, are pushing on a door from opposite sides. Each friend is pushing just as hard as the other but in different directions. If they continue to push in the same direction relative to the door, who is more likely to win? Why?

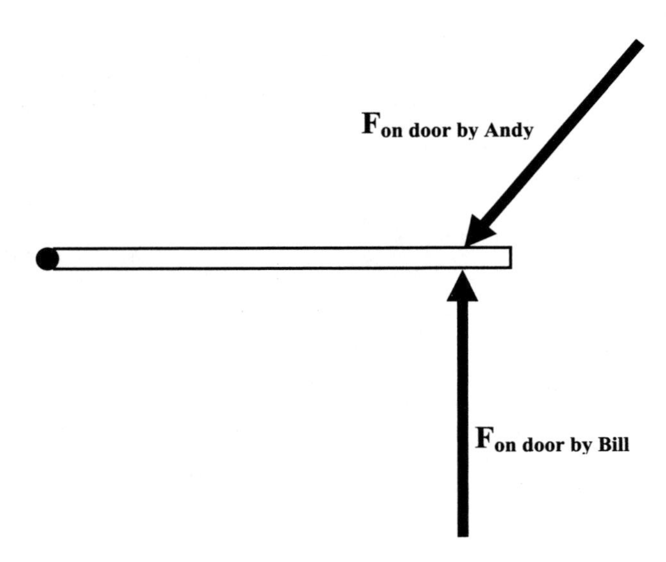

Name ________________________________ Date ____________ Class ____________

a. Does the component parallel to the door contribute to opening or closing the door? Explain.

b. Calculate the components of Andy's force parallel to the door and perpendicular to the door in terms of $\mathbf{F}_{\text{on door by Andy}}$.

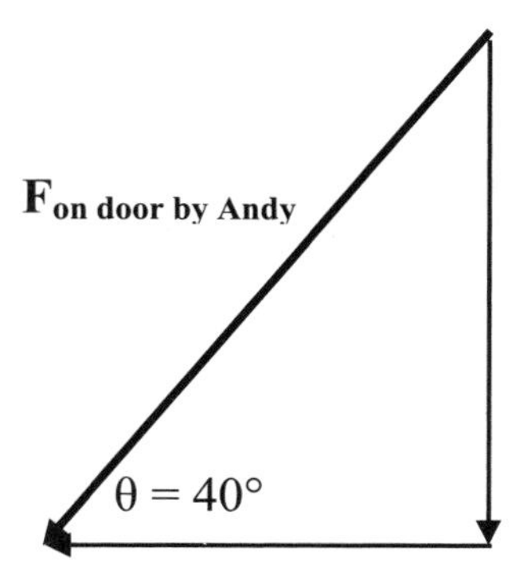

c. Which component opposes Bill's push?

d. How much harder would Andy need to push in order to prevent Bill from moving the door?

e. If Andy were to push on the door closer to the hinge, would he have an easier time opposing Bill? Explain.

f. Both friends are pushing on the door 0.75 m from the hinge. Bill is pushing with 250 N of force. How much torque is he exerting on the door?

g. Andy is pushing back with 350 N (with the same 40° angle). How much of his 350 N is contributing to the torque on the door? (Hint: Which component is opposing Bill?)

Name ______________________ Date __________ Class __________

h. Who manages to move the door? Explain.

23. A block is given a shove and slides along a table, slowing down due to friction between the block and table. The block, initially moving 50 m/s, has a mass of 10 kg and the coefficient of kinetic friction between the block and table is $\mu_k = 0.90$.

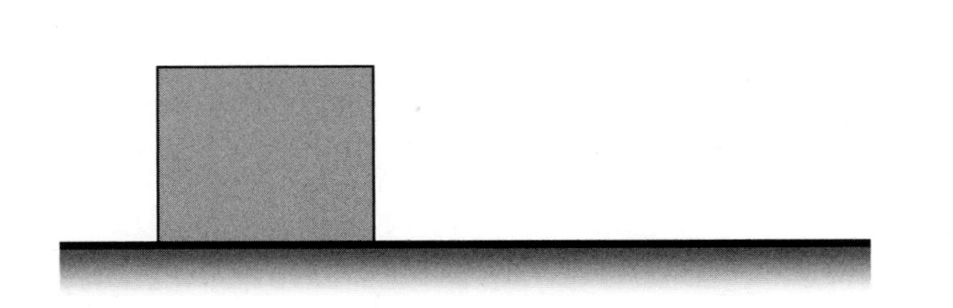

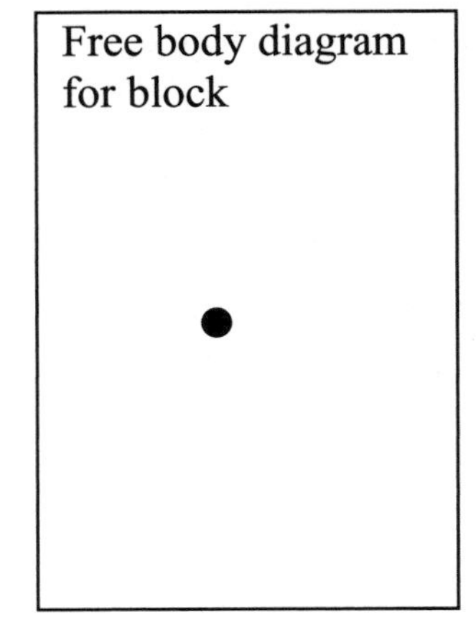

a. Draw the free body diagram for the block. Clearly label each force.

b. Is the block changing speed in the vertical direction? Use Newton's second law to write an equation for the sum of the vertical forces.

c. Find the normal force on the block from the table from the previous equation.

d. Is the block accelerating horizontally? Write an equation for the sum of forces horizontally.

e. Use the previous equation to find the net force experienced by the block.

Name ______________________________ Date ____________ Class ____________

f. What is the acceleration of the block? In which direction is the acceleration vector pointing?

g. Plot v_x versus t and x versus t.

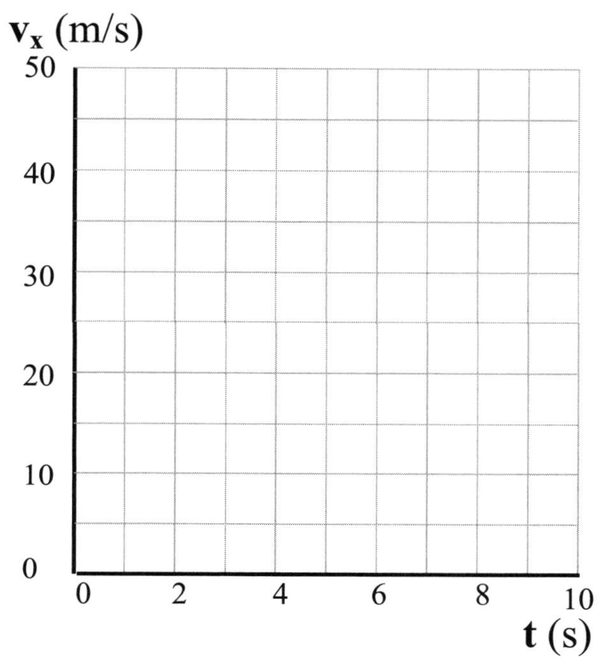

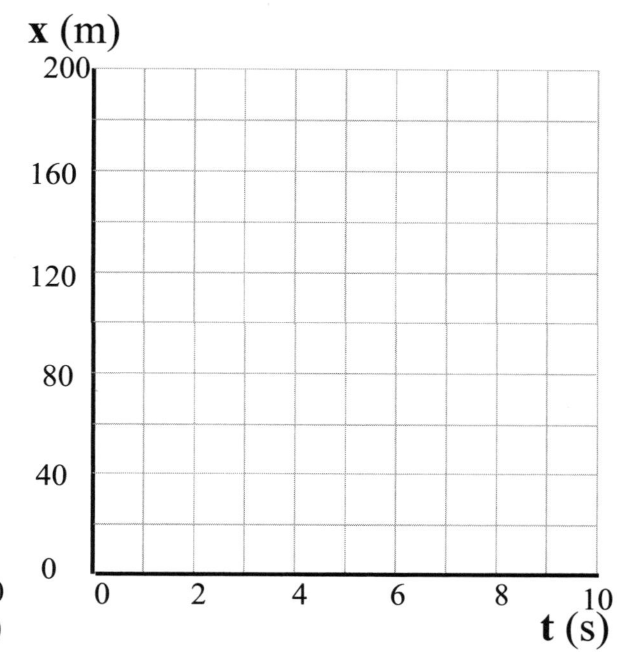

h. How long does it take the block to come to a stop?

i. How far has the block traveled by the time it stops?

Name ______________________________ Date ____________ Class ____________

24. A block is given a shove and slides up a long ramp. The block, initially moving 50 m/s, has a mass of 10 kg and the coefficient of kinetic friction between the block and ramp is $\mu_k = 0.90$.

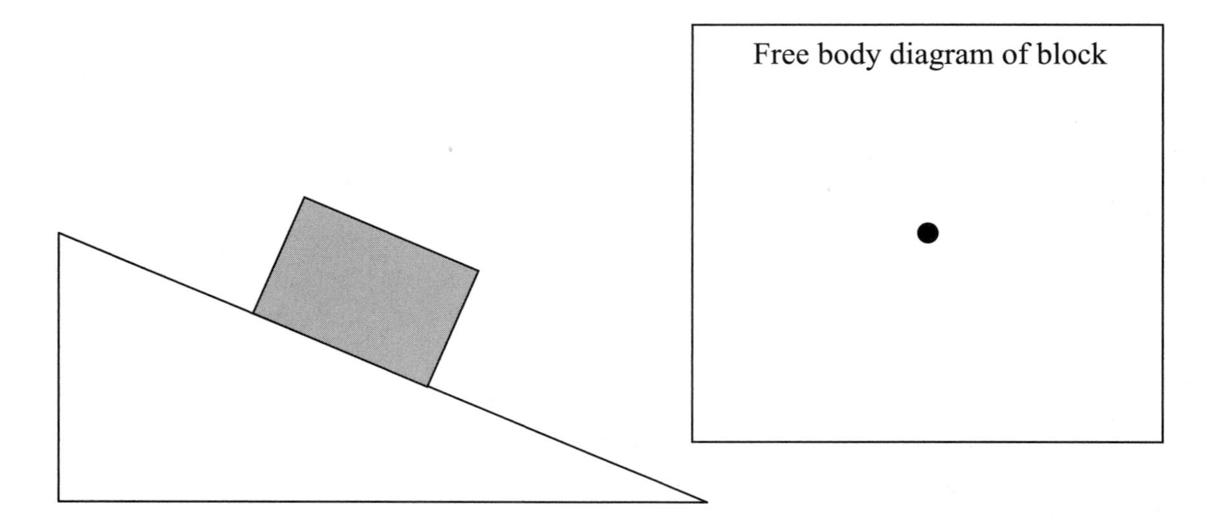

a. What choice of coordinates will make your calculations easier?

b. What is the direction of the normal force on the block?

c. Calculate the components of the weight in your chosen coordinate system.

d. Draw the free body diagram for the block. Clearly label each force.

e. Is the block changing speed in the *y* direction? Use Newton's second law to write an equation for the sum of the forces along the *y*-axis.

f. Find the normal force on the block from the table from the previous equation.

Name ________________________________ Date ____________ Class ____________

g. Is the block accelerating along the x-direction? Write an equation for the sum of forces along the y-axis.

h. Use the previous equation to find the net force experienced by the block.

i. What is the acceleration of the block? Which direction is the acceleration vector pointing?

j. Plot v_x versus t and x versus t.

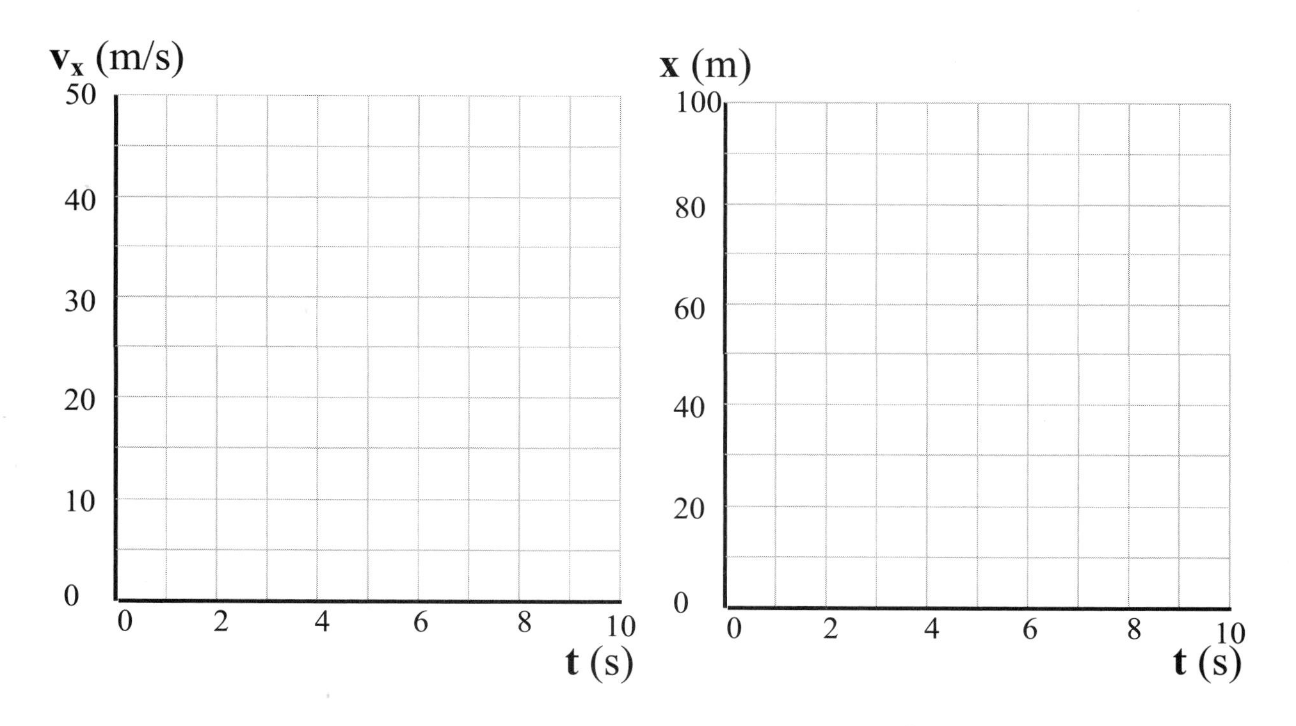

k. How long does it take the block to come to a stop?

l. How far has the block traveled by the time it stops?

Name ______________________________ Date ____________ Class ____________

m. How is this problem different than the previous problem, where the block is sliding on a flat table? What different things did you need to consider for this problem?

n. Write a short description about how you would solve a problem like this in the future.

Name ______________________ Date __________ Class __________

Chapter 6

Bookkeeping on Physical Systems: The Concept of Energy

1. A heavy cart is moving to the right at a constant velocity. For this exercise, you can ignore any friction between the axles and the cart or between the wheels and surface.

   a. If you push on the block in the direction it is moving, what happens to the velocity of the block?

   b. If you push the block to the left, what happens to the velocity of the block?

   c. If you push straight down on the block, what happens to the velocity of the block (ignore friction)?

   d. For each question below, indicate which of the three cases above apply and explain your reasoning.

      i. The energy of the cart increases.

      ii. The energy of the cart decreases.

      iii. You do positive work on the block.

Name ______________________________ Date ____________ Class ____________

iv. You do negative work on the block.

v. You do zero work on the block.

e. Explain to a friend how to tell whether a force is increasing or decreasing the energy of a system and whether the work done on the system is positive or negative.

2. You push on an initially stationary cart (mass = 3 kg) with a constant force of 8 N for 12 s.

a. Calculate the acceleration of the cart. Is the acceleration constant during the 12 s you are pushing?

b. How fast is the cart going after 12 s?

c. How far has the cart moved after 12 s?

d. How much kinetic energy does the cart have after 12 s?

e. The cart was initially stationary. What was its KE at that time?

Name ______________________ Date ____________ Class ____________

f. How much did the energy of the cart change? How was energy transferred to the cart?

g. Since you push on the cart with a constant force over the whole distance it travels, how much work have you done on the system?

h. Why is the work you do on the system equal to the change in the energy of the system?

3. An initially stationary cart starts at the top of a 3.0 m high ramp. You will calculate the velocity at the bottom of the ramp using kinematics and then energy.

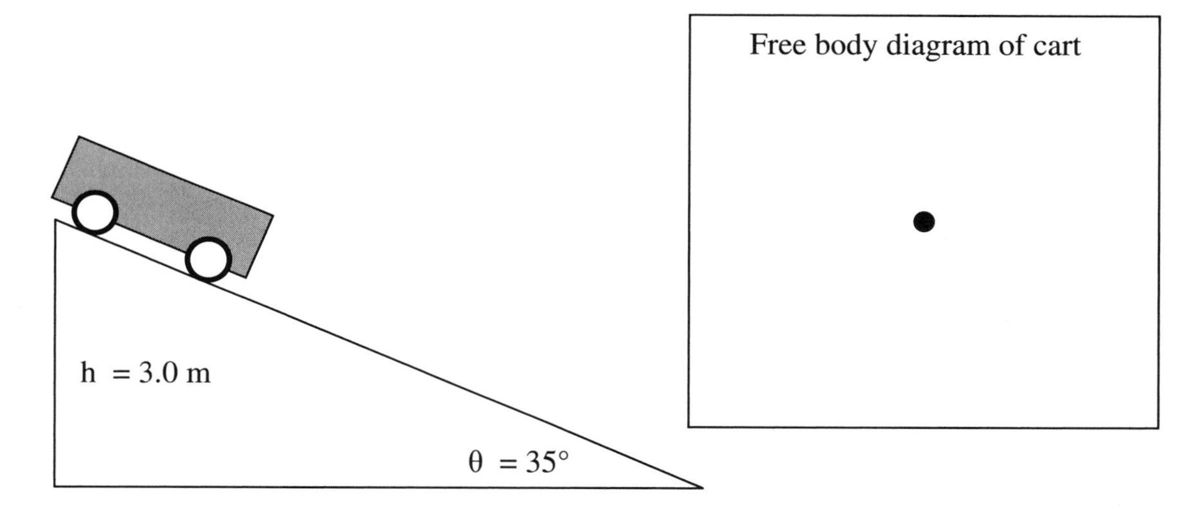

a. Draw the free body diagram for the cart and calculate the net force on the cart.

b. How long is the sloped part of the ramp?

c. The cart has a mass of 1.4 kg. Using the calculations from the two previous questions, find the velocity of the cart at the bottom of the ramp.

4. If the system consists of the cart and Earth, there are two types of energy in the system, gravitational potential energy and kinetic energy. Since the cart is stationary at the top of the ramp it has not kinetic energy, only PE_G.

   a. Calculate the total energy at the top of the ramp. Assume y = 0 m is defined to be the bottom of the ramp.

   b. Since there are no external forces, the system energy is constant.

   c. What is the PE_G of the system when the cart is at the bottom?

   d. Use the kinetic energy of the cart to calculate the velocity at the bottom of the ramp.

   e. If the Earth was not part of our system so it was an external force, how much work would the Earth have done on the cart as it rolled down the entire ramp?

Name ______________________________ Date ____________ Class ____________

f. The same cart is at the top of a curvy ramp starting at the same height. The cart has the same velocity at the bottom of this ramp. Explain in terms of energy why the cart has the same velocity at the bottom of this ramp below as it did with the ramp above.

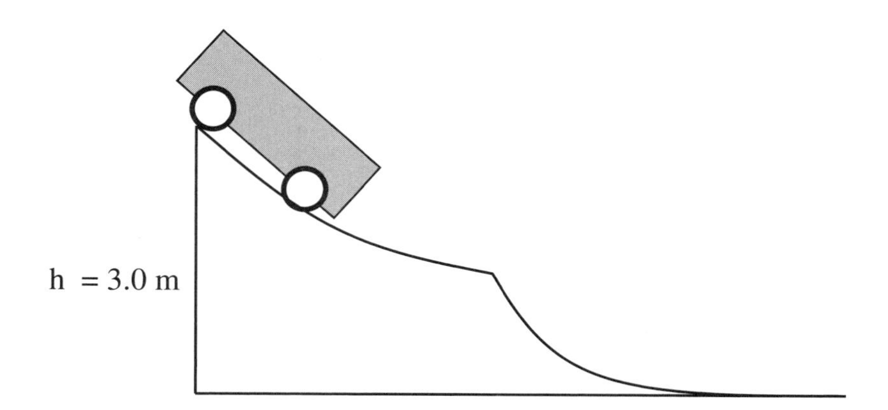

5. A cart is held in place against a compressed spring. When released, the spring pushes the cart towards the right.

a. The system to consider includes the spring and the cart. If the track is level, the Earth doesn't need to be included in the system. Why not?

b. What type of energy does the system initially have?

c. What type of energy does the system have after the cart leaves the spring?

Name ________________________________ Date ____________ Class ____________

d. Fill in the energy bars for each type of energy…

i. before the spring is released

$KE + PE_S = E$

ii. when the spring is partly extended

$KE + PE_S = E$

iii. after the cart has left the spring.

$KE + PE_S = E$

e. If the system only consisted of the cart, what is the initial energy of the system?

f. Does the energy of the system increase or decrease as the spring pushes on the cart?

g. In which direction does the force on the cart point? In which direction does the displacement vector for the cart point?

h. Based on your answer to the previous question, is the work done by the spring on the cart positive, negative, or zero? How do you know?

i. In general, if an external force does positive work on a system, what happens to the total energy of the system?

6. A cart is rolling along a level surface at a constant velocity. At the bottom of the approaching slope is a spring. The cart rolls down the hill, hits the spring, bounces back up the slope, and rolls along the level surface with a constant velocity. For this exercise you can treat any dissipative energy losses as negligibly small.

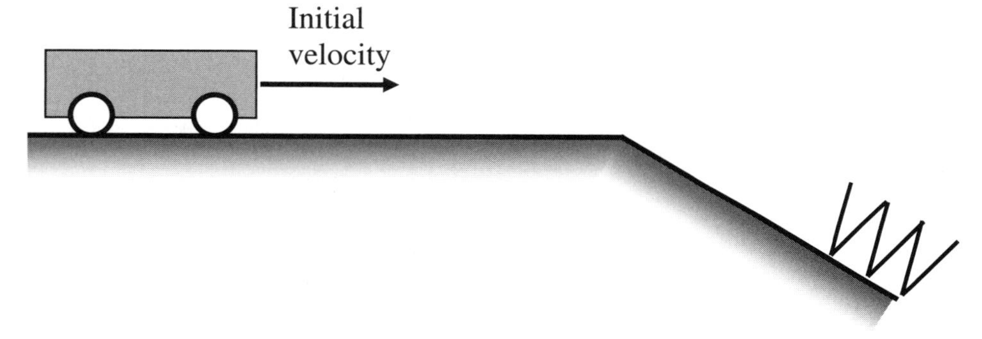

a. There are several different choices of systems for this problem. For each system, list the types of energy the system contains and which external forces do work on the system.

i. System 1: The cart, spring, and Earth

ii. System 2: The cart and the spring

iii. System 3: The cart and Earth

iv. System 4: Only the cart

b. The total energy for System 1 is constant (calculating the total energy for the system at any time gives the same value). Why isn't the total energy constant for the other three systems?

c. Consider System 3, which contains the cart and Earth. Is the spring doing positive work, negative work, or no work on the system as the spring is compressed? Explain how you can tell.

d. What is happening to the total energy of System 2 as the spring is compressed?

e. What happens to the total energy of System 2 as the spring starts to extend? Is the work done by the spring on the system positive or negative?

Name ______________________________ Date ____________ Class ____________

f. Fill in the energy bars at the points indicated in the picture if you are considering System 1. Point B is halfway between the spring and level area. The spring is compressed the most at point C and the gravitational potential energy is zero at point C.

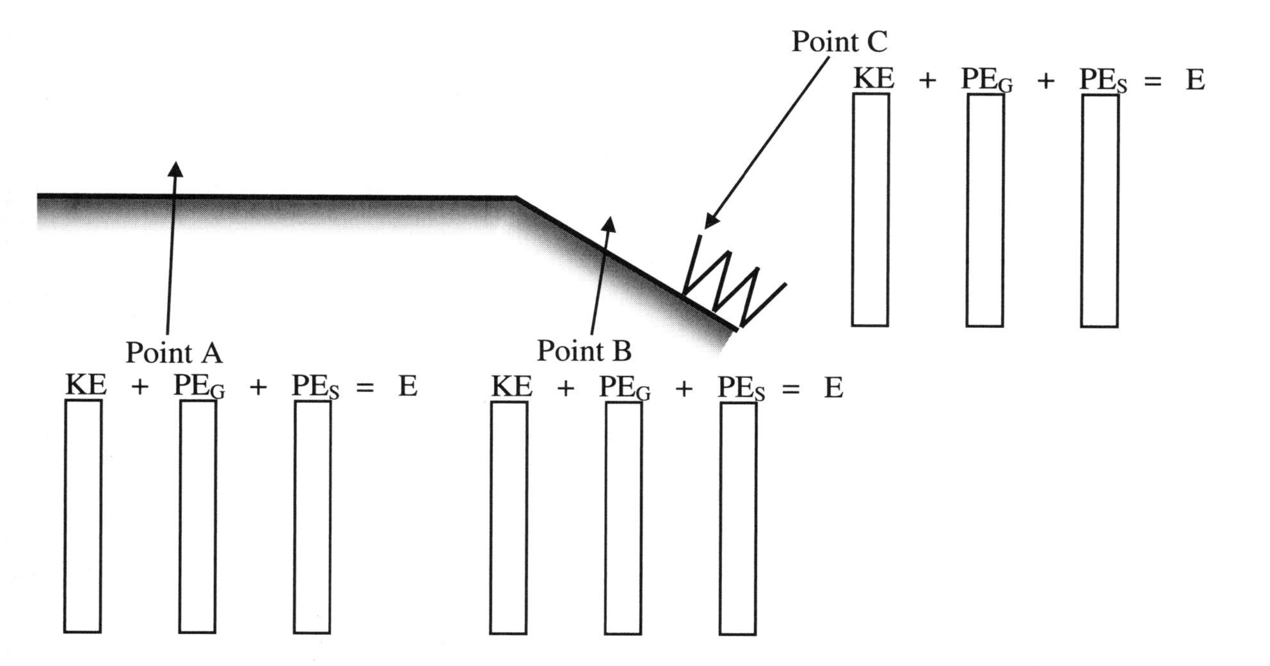

The cart has a mass of 2.0 kg and is traveling at 1.7 m/s at point A. The spring constant for the spring is k = 14 N/m and the maximum change in the cart's height is 2.2 m.

g. Calculate the kinetic and potential energy at point A and find the total energy E. Compare your calculations to the energy bars above and make any corrections to the bars.

h. Calculate KE, PE, and E at point B (the height at B is 1.1 m). Compare your answers to the bar graphs at B.

i. Calculate KE, PE, and E at point C. Compare your results to the energy bars.

j. If the spring is replaced with another spring twice as stiff, how does the total energy of System 1 change? Explain your reasoning.

k. If the steepness of the slope is increased (but the change in height is the same as before), how does the total energy of System 1 change? Explain your reasoning.

l. If the spring is moved farther down the slope, how does the total energy of System 1 change?

Name ______________________ Date __________ Class __________

Chapter 7

More Bookkeeping: Collisions and the Concept of Momentum

1. In problem 2 of the previous chapter, you found the final velocity of a cart pushed for 12 s with a force of 8 N. The cart was initially stationary and had a mass of 3 kg.

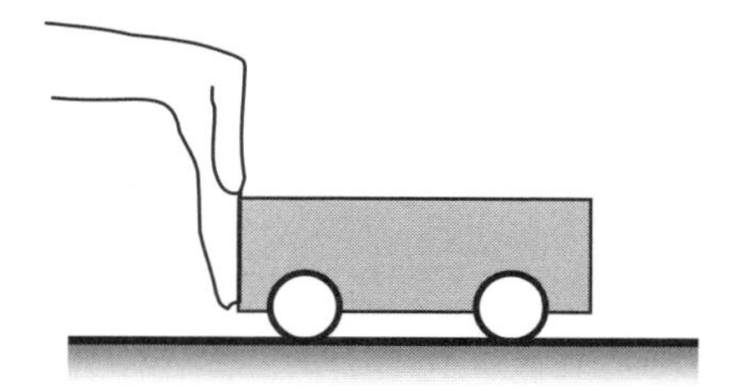

   a. Calculate the impulse delivered to the cart by the hand over the entire 12 s.

   b. What is the initial momentum of the cart?

   c. The impulse tells you how much the momentum has changed. Use the impulse and initial momentum of the cart to find its final momentum.

   d. Calculate the final velocity using the momentum calculated in the previous question. Is this value the same answer you calculated in problem 2 of the previous chapter?

e. Of the three different approaches, kinematic, work and energy, and momentum transfer, which was the easiest approach for calculating the final velocity? Explain your reasoning.

2. Two identical cars moving at 14.0 m/s approach a red stop light. The driver of car A gently applies the brake while the driver in Car B waits until the last moment to slam on the brakes. (Assume the force of the brakes is constant for both cars.)

   a. Do you expect one of the cars to experience a larger braking force or will the braking force be the same for both cars? Explain your reasons for choosing your answer.

   b. Which car do you think undergoes a larger change in momentum? Explain your reasons.

   c. Will either car experience a larger impulse or will they both have the same impulse? Explain your reasoning.

   d. Each car has a mass of 1600 kg. Car A stops in 6 s and car B takes 2 s to stop.

   e. Calculate the initial momentum of each car.

   f. What is the final momentum of the cars? Find the total change in momentum for each car.

   g. What is the impulse exerted on each car?

h. Use the value for the impulse to find the average braking force each car experiences.

i. Were your predictions in ***a***, ***b***, and ***c*** correct? If not, explain why your predictions were off.

j. What other piece of information would you need to calculate the energy transfer between the car and road? Explain how you would calculate this information and how you would use it to find the energy transfer (do not do the calculations).

3. Two friends, Michelle and Neal, are discussing a collision where a moving truck hits a stationary car. Michelle says, "Even though the truck is more massive, it pushes on the car just as hard as the car pushes back, so the momentum gained by the car must be equal to the momentum lost by the truck." Neal says, "But the truck is heavier, so it will lose less momentum than the car gains."

   a. Which of Newton's laws back up what Michelle said?

   b. What is wrong with Neal's statement?

   c. Neal also states, "The car changes speed more than the truck, so the car must have gained more momentum than the truck." Explain to Neal why he is wrong.

Name ______________________________ Date ____________ Class ____________

4. State whether each collision is best treated as elastic, totally inelastic, or partially inelastic. Explain your reasoning.

   a. A pool ball striking another pool ball.

   b. A rubber ball striking the floor and bouncing up, but not as high as the previous bounce.

   c. A train car slamming into another train car as they are coupled together.

   d. Two cars colliding.

   e. Two cars colliding (including the Earth in the system).

Name ______________________________ **Date** ____________ **Class** ____________

5. A stopped car is rear-ended by a large truck. Assume the collision is perfectly elastic and any external forces are negligible. Rank in order, from largest to smallest, the final velocities of the car.

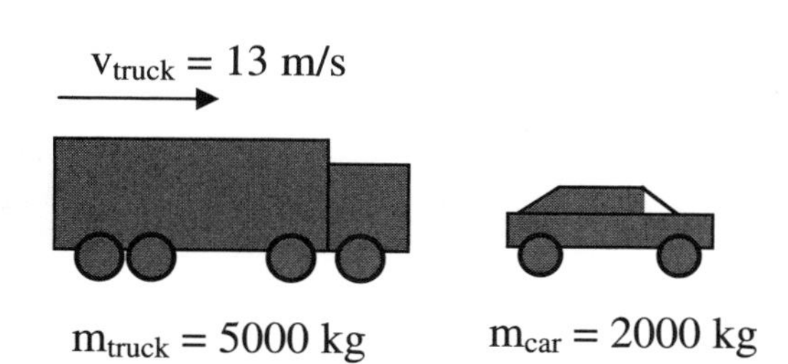

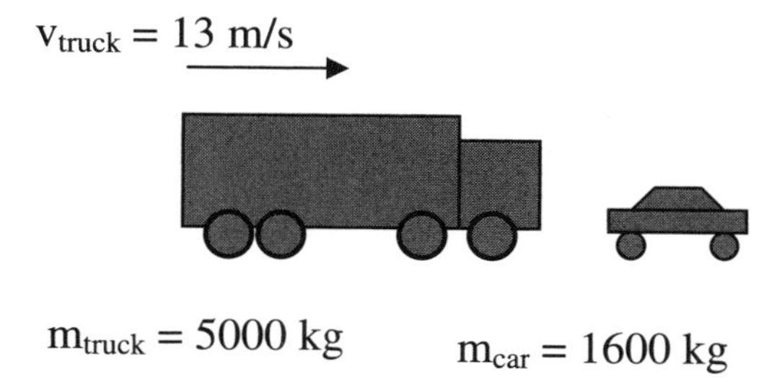

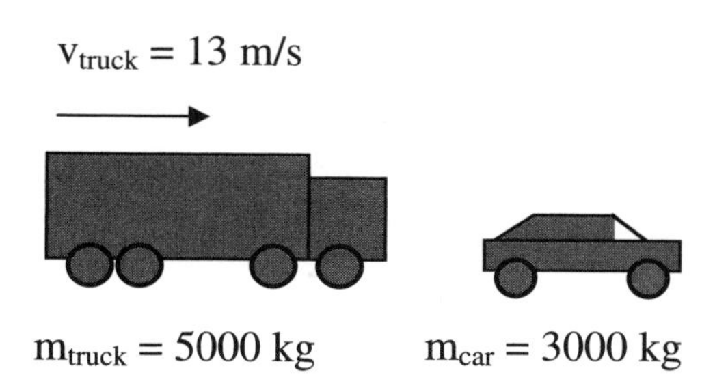

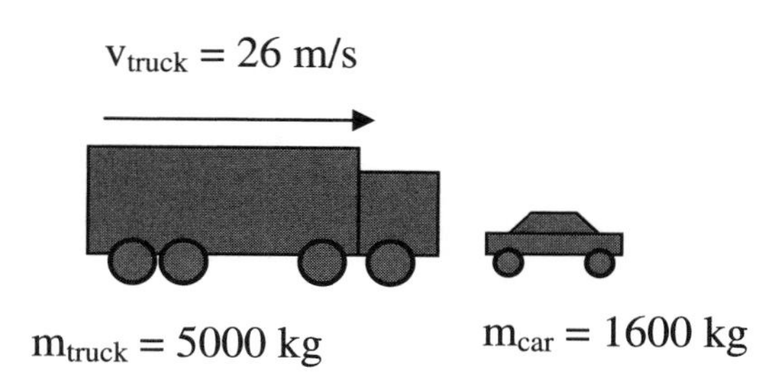

6. If the collision between the truck and car in the previous problem had been totally inelastic, would the final velocity of the car be greater than, the same as, or smaller than the final velocities for an elastic collision? Explain your reasoning.

7. A force sensor mounted on cart A records an average force of F for Δt the cart runs into stationary cart B.

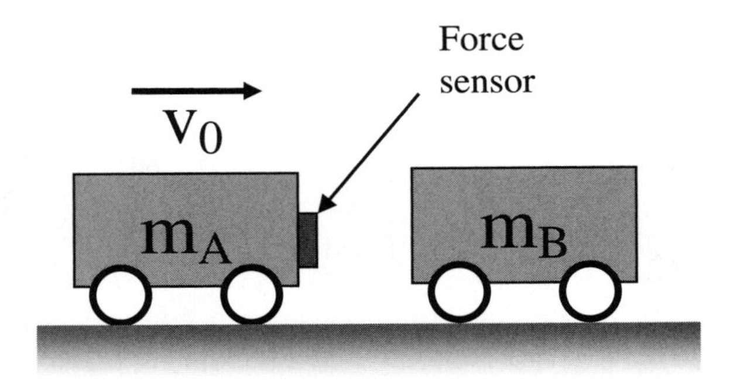

a. How should the force F recorded by the sensor be labeled: $F_{\text{on A by B}}$, $F_{\text{on B by A}}$, or something else? Explain your answer.

b. Write the equation for the average acceleration cart B experiences during the collision in terms of the average force cart B experiences. In which direction does B accelerate?

c. Write the equation for the velocity of B after the collision in terms of the average force F, the time interval of the collision Δt, and the mass of the cart.

d. What is the average force cart A experiences as a result of the collision?

e. What is the direction of the force exerted on A?

f. In terms of force F and the time interval of the collision Δt, write the equation for the final velocity of cart A. Keep in mind that cart A has an initial velocity.

Name ______________________________ Date ____________ Class ____________

g. Express the change in momentum of cart A in terms of the average force F and Δt.

8. A heavy steel ball is hanging by a cable inside a cart that is free to move. Initially the ball is pulled back and held in place by a rope. Assume friction is negligible.

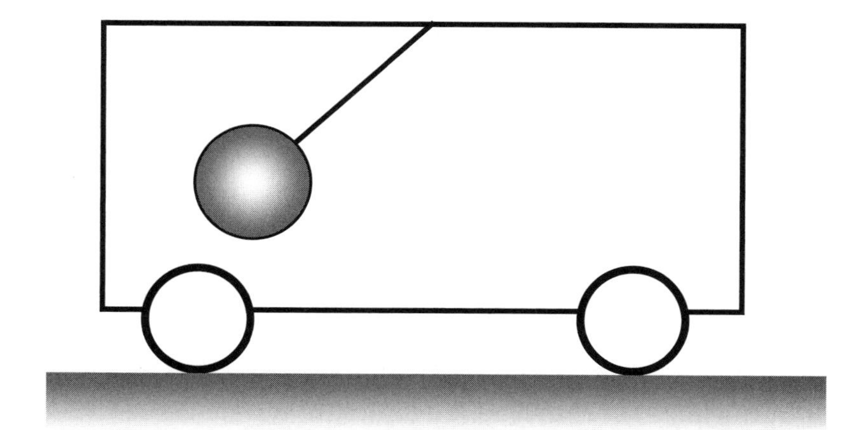

a. Neither the ball nor the cart is moving initially. What is the total momentum of the cart and ball system?

b. The rope holding the ball up breaks and it swings back and forth on the cable.

c. There are no forces acting on the ball and cart system horizontally. Does this mean the horizontal component of momentum is conserved? Explain.

d. What two forces act on the ball and cart system vertically?

e. Is the vertical component of momentum conserved? Explain.

Name ______________________ Date ____________ Class ____________

f. In the area below, draw a free body diagram for the ball at the top of swing to the left, at the bottom of the swing, and at the top of the swing to the right. Do the same for the cart for those three cases. Assume all the forces exerted on the cart can be treated as acting on one point. Label all forces and draw the net force.

(1) Top of swing left

Free body diagram for ball	Free body diagram for cart
•	•

(2) Bottom of swing

Free body diagram for ball	Free body diagram for cart
•	•

(3) Top of swing right

Free body diagram for ball	Free body diagram for cart
•	•

Name ______________________________ Date ____________ Class ____________

g. The cart moves back and forth as the ball swings. What is exerting the force on the cart that pushes it back and forth?

h. When is the force experienced by the cart the largest?

i. At what point in the swing is the momentum of the ball largest? Explain your reasoning.

j. The momentum of the cart is largest at the very bottom of the ball's swing. Is the total momentum of the ball and cart system still zero at the bottom of the swing? Explain your answer.

k. The cart and ball always move in opposite directions at the bottom of the swing. Explain in terms of the total momentum of the system why this must be the case.

l. Is the momentum of the cart equal but opposite to the ball at the bottom of the swing? Explain.

m. Why do the cart and ball have different velocities at the bottom of the swing?

Name ______________________________ Date ____________ Class ____________

n. Why isn't total energy conserved for the cart and ball system? Explain your reasons.

o. If the system consists of the ball, the cart, and the Earth, is the total energy of the system conserved? Explain.

p. What types of energy does the system of the ball, the cart, and the Earth have?

q. Where is the kinetic energy of the ball the greatest?

r. At the bottom of the swing, the cart and ball do not have the same kinetic energy. Why not?

s. Sketch the energy bars for the system at the top and bottom of the swing.

t. The cart has a mass of 700 kg and the ball is 300 kg. The ball is initially lifted 1.5 m from the bottom of the swing.

u. Calculate the initial momentum of the ball, the cart, and the system containing the ball and cart.

v. Without using energy conservation, is there an easy way to find the velocity of the ball and cart at the bottom of the swing?

w. Calculate the initial total energy of the system (i.e. when the cart and ball are initially stationary).

x. Calculate the kinetic energy of the ball and cart at the bottom of the swing. Use these values to find the velocity for each.

y. Calculate the momentum of the cart and the ball at the bottom of the swing.

Name ______________________ **Date** ____________ **Class** ____________

z. Write a brief summary of the steps you took to find the momentum of the cart and ball at the lowest point of the ball's arc.

Name ______________________ Date __________ Class __________

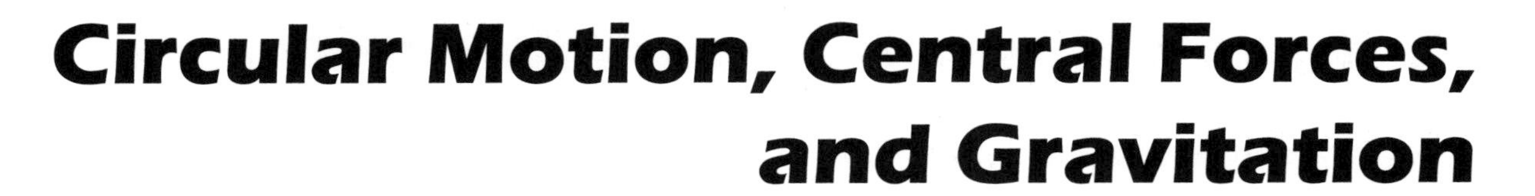

Circular Motion, Central Forces, and Gravitation

1. A physics instructor is rollerblading down a steep hill. This is his first time out, so he doesn't know how to brake. He reaches level ground but he doesn't slow down much. He passes a sign post which he manages to barely catch with his right hand.

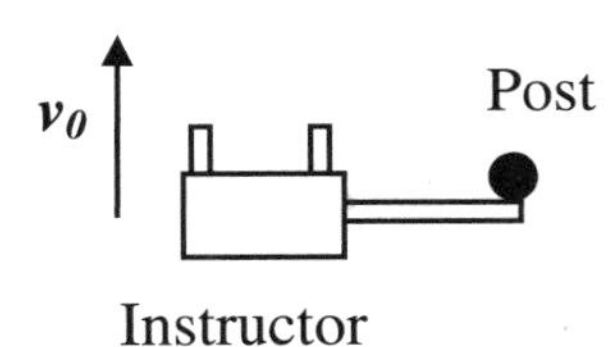

   a. Draw the free body diagram for the hapless instructor just after he catches the sign post. Draw the diagram from above looking down (i.e. do not include weight and normal force).

Name ______________________ Date __________ Class __________

b. Draw a free body diagram for the post just after he catches it. Draw the diagram from above looking down. Identify the third law pair to the force exerted by the instructor on the sign post.

Free body diagram for instructor	Free body diagram for post
•	•

c. Draw the free body diagram after he has moved through a quarter of the circle. Assume he hasn't slowed down much yet.

d. Draw a free body diagram for the post at this time. Draw the diagram from above looking down. Identify the third law pair to the force exerted by the physics instructor on the sign post.

Free body diagram for instructor	Free body diagram for post
•	•

Name ______________________________ Date ____________ Class ____________

e. Why has the force the post exerted on the instructor changed direction? What effect does this have on his motion?

2. A ball is traveling clockwise in a circle at a constant speed.

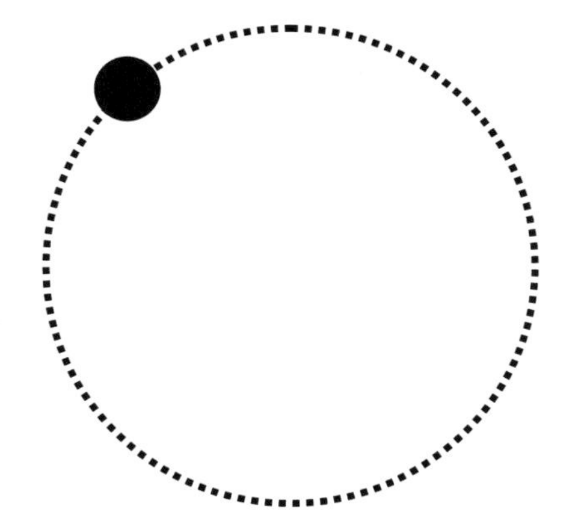

a. Draw the net force exerted on the ball on the diagram above.

b. If the ball was traveling in the opposite direction, would the net force on the ball be different? If yes, draw the force on the diagram. Explain your reasons for your answer.

c. Explain how the force exerted on the ball would or would not change for each of the following cases.

i. Increasing the mass of the ball.

ii. Decreasing the size of the ball without changing the mass.

iii. Increasing the radius of the circle.

Name ______________________________ Date ____________ Class ____________

iv. Increasing the speed of the ball.

3. You are pushing a girl standing on a merry-go-round with a radius of 2.2 m. The girl is sitting 1.4 m from the center. The merry-go-round goes around once each 8 s.

a. How far does the girl travel each revolution?

b. What is the magnitude of her tangential velocity?

c. The floor is very slick so she has to hang on to the rail to keep herself from flying off. How hard does she have to grip the railing?

d. How hard would she need to pull to move until she is only 1.0 m from the center?

Name ________________ Date ________ Class ________

4. A ball moves along an inwardly spiraling track. Assume friction is negligible and no other forces besides those from the track are exerted on the ball.

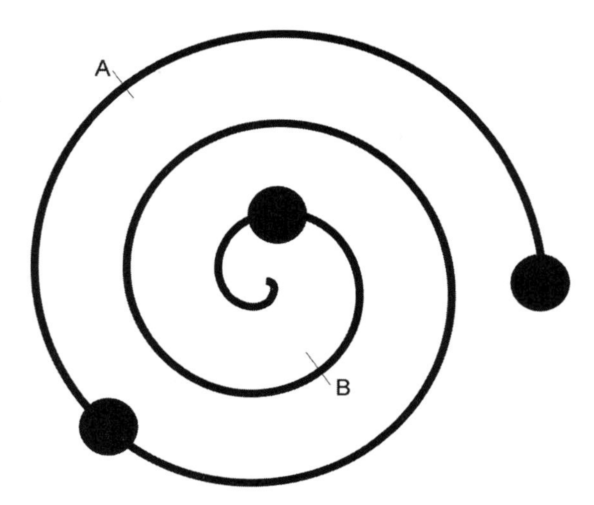

a. The direction of the tangential velocity changes as the ball goes around. Does the magnitude of the tangential velocity change as the ball spirals in? Explain your answer.

b. The direction of the radial force exerted on an object changes as the object follows a circular path. Does the magnitude of the radial force change as the ball spirals inward? Explain your answer and draw the force vectors on the ball at the three positions in the earlier diagram.

c. If the ball started near the center and was spiraling outward, explain how the radial force experienced by the ball would behave. Draw the radial force vectors at the three positions in the earlier diagram.

d. Sketch the path the ball would take if the track was cut along line **A**.

Name ______________________________ Date ___________ Class ___________

e. Sketch the path the ball would take if the track was cut along line **B**.

5. The space shuttle is moving in a stable circular orbit around Earth.

   a. Sketch the free body diagram for the shuttle.

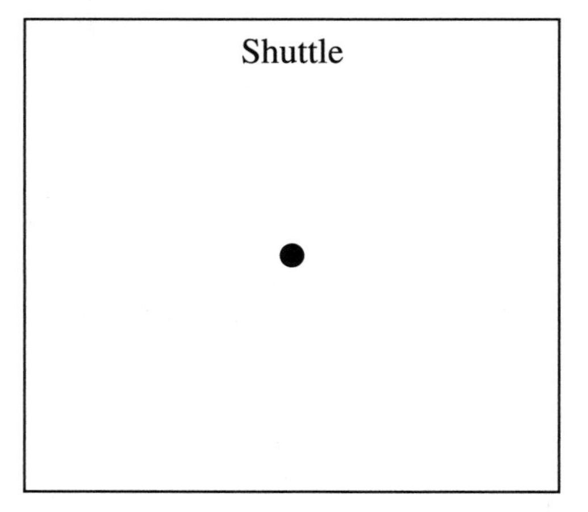

   b. How does the magnitude of the force of gravity exerted by Earth on the space shuttle compare to the magnitude of the centripetal force needed to keep the space shuttle in a circular orbit?

   c. If the shuttle briefly fires its engines tangential to its path and speeds up, will the shuttle remain in the same orbit? Explain your answer.

   d. Can the shuttle stay in the same circular orbit while moving at a faster speed? If you think it can, explain how. If you don't think it can, give your reasons.

Name ______________________________ Date ____________ Class ____________

e. Calculate the gravitational force exerted by the Earth on the space shuttle if it is orbiting 300 km above the surface of the Earth. (The mass of the space shuttle is roughly 10^5 kg.)

f. How much does a 75 kg astronaut weigh on the ground?

g. Calculate the gravitational force exerted by the Earth on an astronaut aboard the space shuttle. What percent of their weight on the ground is this?

h. How much force does the orbiting space shuttle exert on the Earth?

i. How much force does the astronaut exert on the Earth while in orbit?

Name ______________________________ Date ____________ Class ____________

Chapter 9

Rotational Kinematics and Dynamics

1. An electric toy train moves along a track at a constant speed. It takes four seconds for the train to travel from one end of the track to the other.

a. Sketch the x versus t graph for the train.

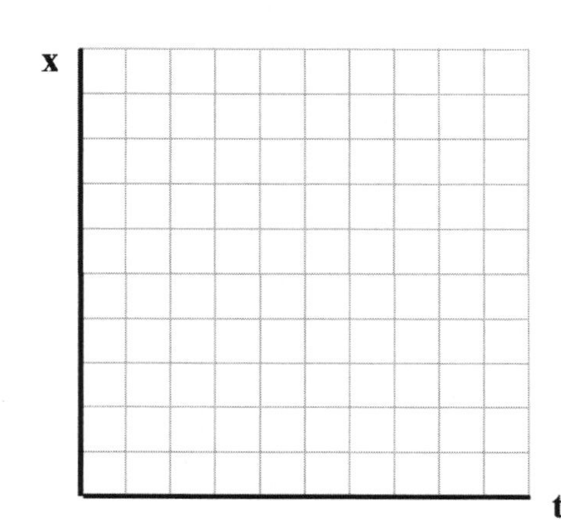

b. The track is reshaped into a circle. Sketch the distance traveled versus clock time for the train.

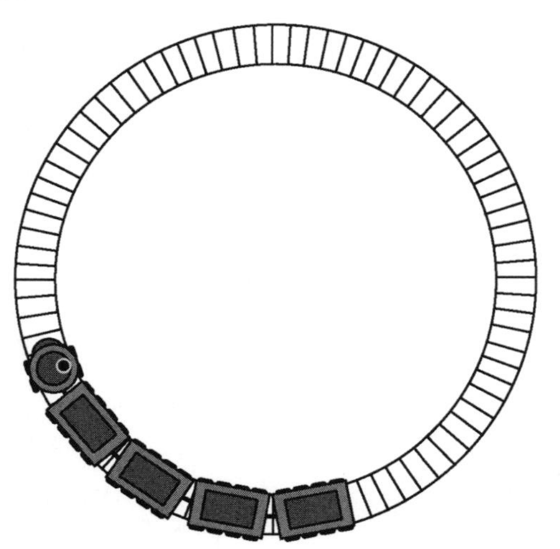

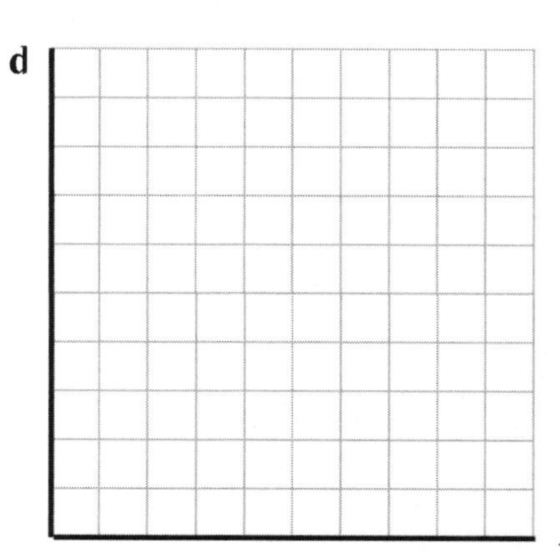

Name ______________________________ Date ____________ Class ____________

c. Sketch the angular position of the train versus clock time.

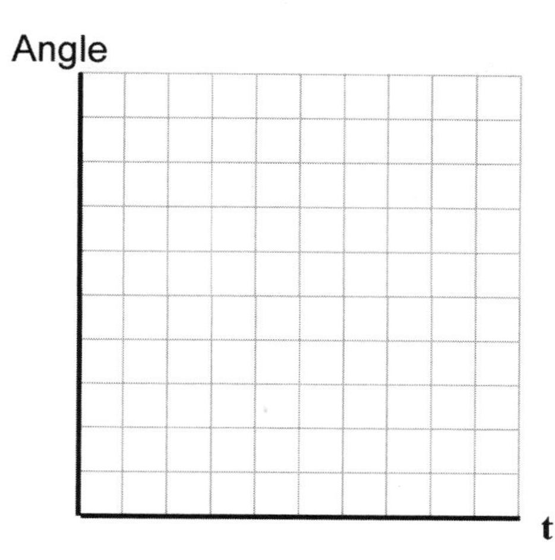

d. Is the linear velocity of the train constant? Is the angular velocity constant?

e. Can you calculate the angular velocity without knowing the track length? If yes, then calculate the angular velocity. If no, then explain why not.

The track is laid out in a spiral and the train is moving clockwise.

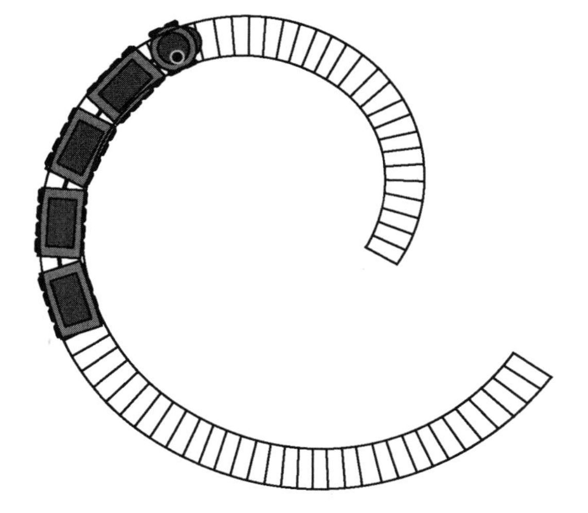

f. How does the linear speed of the train change as it spirals in?

Name ______________________________ Date ____________ Class ____________

g. How does the angular speed of the train change as it spirals in?

h. Sketch the distance traveled and the angular position of the train as it spirals inward.

i. Explain to a friend how angular speed can change without the linear speed changing.

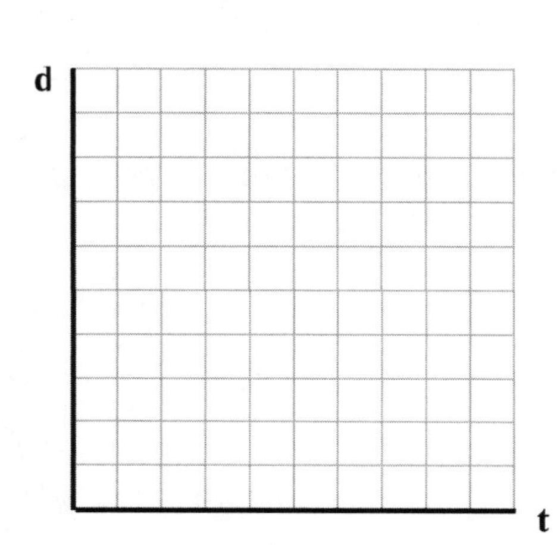

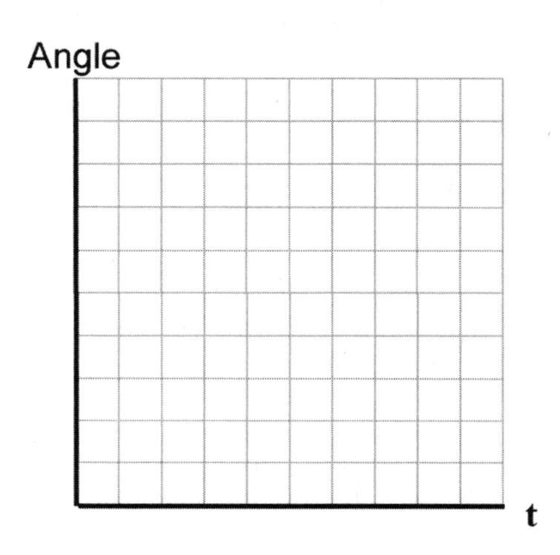

2. Strobe pictures of two objects moving clockwise in a circle at different speeds are shown below. The numbers represent the clock reading in s.

a. How can you tell which object is moving faster?

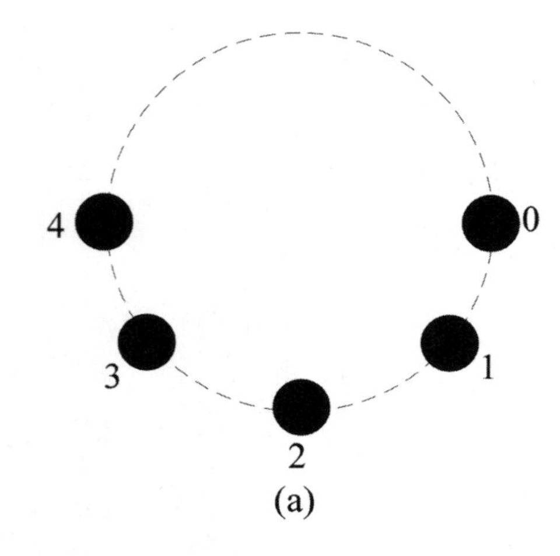

(a)

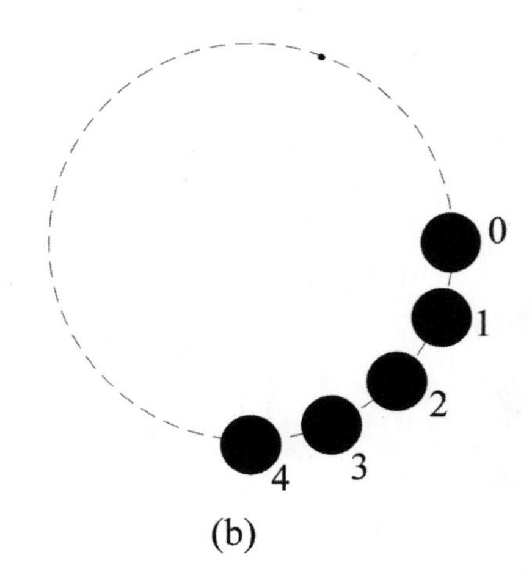

(b)

Name ______________________ Date ____________ Class ____________

b. The angle versus clock reading graph below is for (a). Sketch the angle versus clock reading graph for (b) on the same graph.

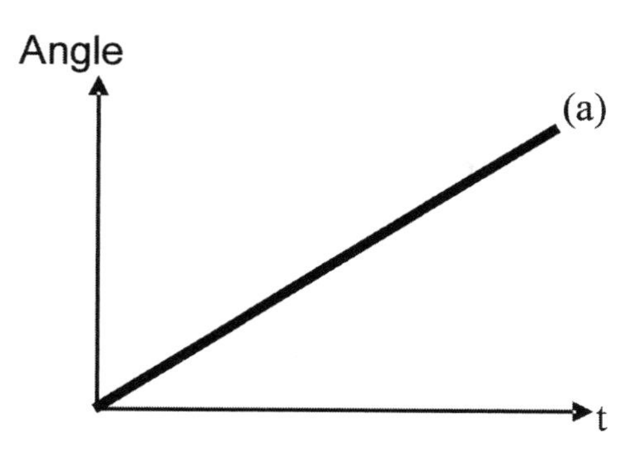

c. Write the equation relating angular position θ with arc length s.

d. The radius of the circular path for both balls is 3.0 cm. Calculate the distance traveled (arc length) s by each ball after 2 s, 4 s, 6 s, and 8 s and then draw the graph of distance traveled s versus t for both balls.

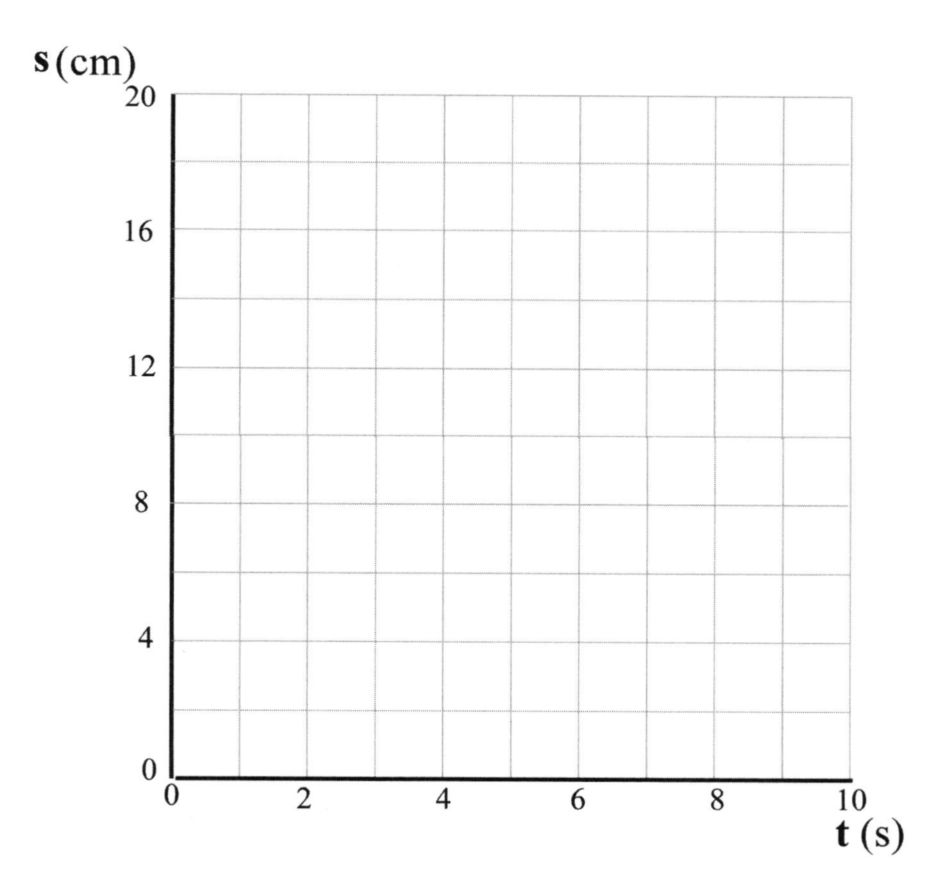

Name ______________________________ Date ____________ Class ____________

e. Calculate the linear velocity of both balls.

f. Calculate the angular velocity for both balls.

g. If the radius of the circular path was 4.0 cm, how would the linear and angular velocity change?

h. Calculate the linear and angular velocity for the 4.0 cm path.

i. Does angular velocity depend on the size of a circle?

3. You are pushing a girl standing on a merry-go-round with a radius of 2.2 m. The girl is holding on to a rail 1.4 m from the center. The merry-go-round is initially stationary, but after pushing for 10 s, the girl is going around once a second.

Name ______________________________ Date ____________ Class ____________

a. Draw a free body diagram of the girl at a particular instant. Assume all the forces on her act at a single point and treat her like a solid object without structure. Indicate the direction the girl is moving and the direction to the center of the merry-go-round.

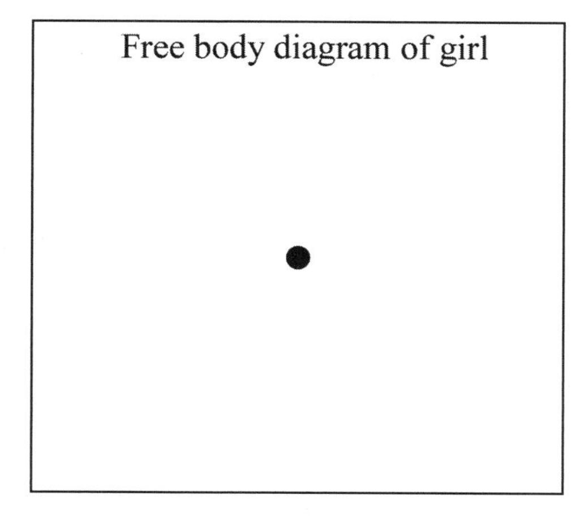

b. What was her average angular acceleration while you were pushing her?

c. What is her final angular acceleration?

d. Did the girl undergo constant linear acceleration? Explain your reasoning.

e. If the girl's mass is 35 kg, how much force is exerted on her while the merry-go-round was speeding up? What is responsible for this force and which way is it pushing on her?

f. Calculate the torque about the center of the merry-go-round experienced by the girl.

g. What other piece of information about the merry-go-round would you need if you wanted to calculate the torque you exerted as you turned the merry-go-round?

h. If she had been sitting at the edge, how much force is exerted on her as the merry-go-round accelerates?

i. Noticing how much easier it is to push the merry-go-round when the girl is close to the center, a friend tells her to sit in the middle while he gets the merry-go-round moving fast. The friend figures she can then move out toward the edge while the merry-go-round keeps spinning at the same speed. What is wrong with this idea? Explain what will happen as she walks toward the edge.

j. As the merry-go-round is spinning, the girl tries to pull herself toward the middle. Why is it so difficult for her to pull herself to the very middle of the merry-go-round?

k. As she moves toward the center, why does the merry-go-round speed up?

l. As she moves toward the center, does her tangential speed increase?

Name ______________________________ Date ____________ Class ____________

4. Two identical balls with mass m are launched with the same initial velocity v_0. Assume the balls do not slow down as they go around the tracks.

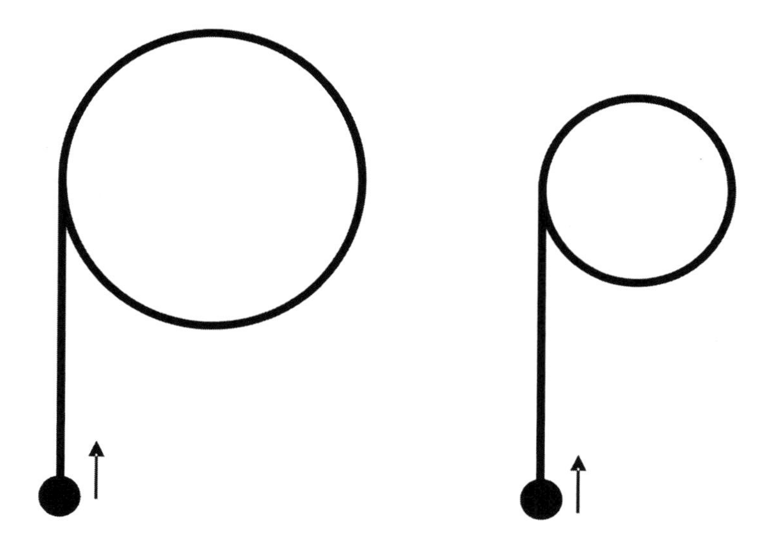

a. What is the initial linear momentum of each ball in terms of m and v_0?

b. Once the balls start going around, the direction of the linear momentum changes. Does the magnitude of the linear momentum change as the balls go around?

c. Do the balls have the same angular momentum about the center of each track? Explain your answer.

d. Calculate the angular momentum for each ball.

e. Explain how each of the following would affect the angular momentum of the balls about the center of the track.

i. Increasing the length of the straight part of the track.

Name ______________________ Date __________ Class __________

ii. Decreasing the mass of the balls.

iii. Decreasing the initial velocity of the balls.

iv. Increasing the diameter of the track

5. A ball moves along an inwardly spiraling track. Assume friction is negligible and no other forces besides those from the track are exerted on the ball.

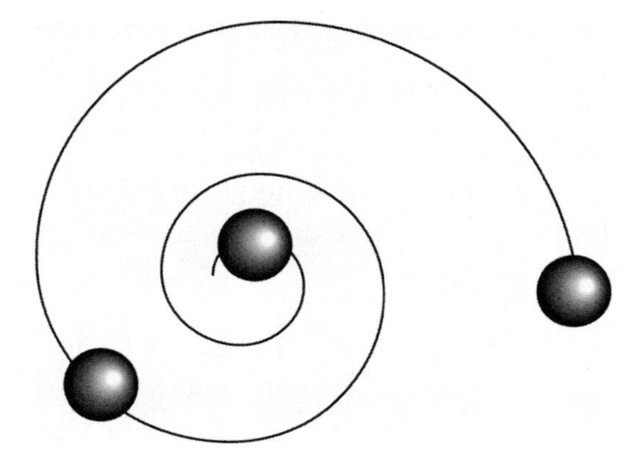

a. Does the spiral exert a tangential force or only a radial force on the ball as it rolls around the spiral?

b. The direction of the tangential velocity changes as the ball goes around. Does the magnitude of the tangential velocity change as the ball spirals in? Explain your answer.

c. Does the angular velocity of the ball change as the ball spirals in? Explain your answer.

d. The direction of the radial force exerted on an object changes as the object follows a circular path. Does the magnitude of the radial force change as the ball spirals inward? Explain your answer and draw the force vectors on the ball at the three positions in the diagram above.

e. If the ball started near the center and was spiraling outward, explain how the radial force experienced by the ball would behave. Draw the net force vectors at the three positions in the diagram above.

f. Sketch the magnitude of the force exerted by the track on the ball as a function of the position.

g. Does the track do work on the ball as the ball rolls around the spiral?

Name ______________________ Date ____________ Class ____________

h. In light of your answer to the previous question, what happens to the kinetic energy of the ball as it spirals inward?

i. The kinetic energy of the ball depends on the velocity of the ball. Does your answer to the previous question agree with your answer to ***b***? Explain why or why not.

Name ______________________________ Date ____________ Class ____________

Chapter 10

Statics and Dynamics of Fluids

1. Atmospheric pressure is usually around 14.7 lb/in^2. If you were to lift an 8 1/2" × 11" sheet of paper off the table, which way is air pressure pushing on the paper? Explain your answer.

   a. Calculate the surface area of the sheet of paper in square feet.

   b. Calculate the downward force on the surface of the paper due to air pressure. Convert this number into tons.

   c. Why doesn't the piece of paper rip if there is that much force pushing down on it?

   d. Explain why pressure has no direction.

Name ______________________ Date ____________ Class ____________

2. A piston is used to seal a cylindrical container. The piston does not leak but is free to move inside the cylinder. The piston is light enough that its weight is negligible.

   a. A pin is put in place to prevent the piston from moving. The container is submerged a fair distance. Draw the two forces experienced by the pistons and the net force immediately after the pin is removed.

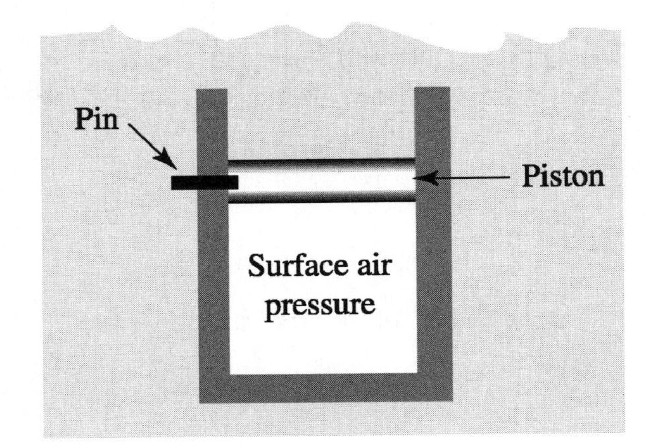

   b. Which way does the piston move? Sketch the new location of the piston on the previous diagram.

   c. How can you tell that the net force experienced by the piston is 0 N?

   d. Draw the two forces experienced by the piston.

Name ______________________________ Date ____________ Class ____________

e. The container is flipped over, keeping the piston at the same depth so the water pressure does not change. Explain why the piston does not move.

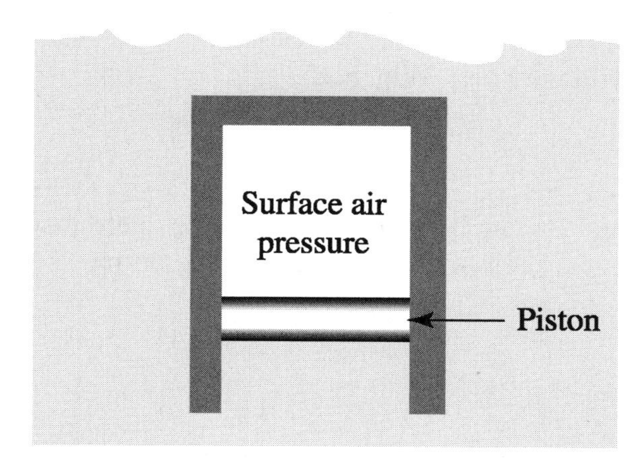

f. The device is flipped on its side. Does the force exerted on the piston due to the pressure difference change at all? Explain why or why not.

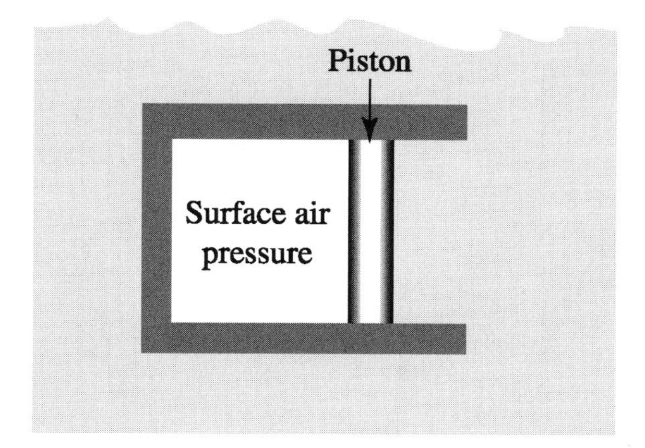

Name ______________________________ Date ____________ Class ____________

3. An opaque bottle is filled with water and then the top is sealed. After punching a small hole in the side the water flows but eventually comes to a stop.

   a. Even though no more water is flowing out of the hole, you cannot say for sure that the water line is even with the hole. Why not?

   b. Once the water stops flowing, how does the water pressure just inside the hole compare with the surrounding air pressure? Explain your reasoning.

4. A flat, thin block with a mass of 2 kg is submerged in water. The top of the block is 0.5 m below the surface of the water.

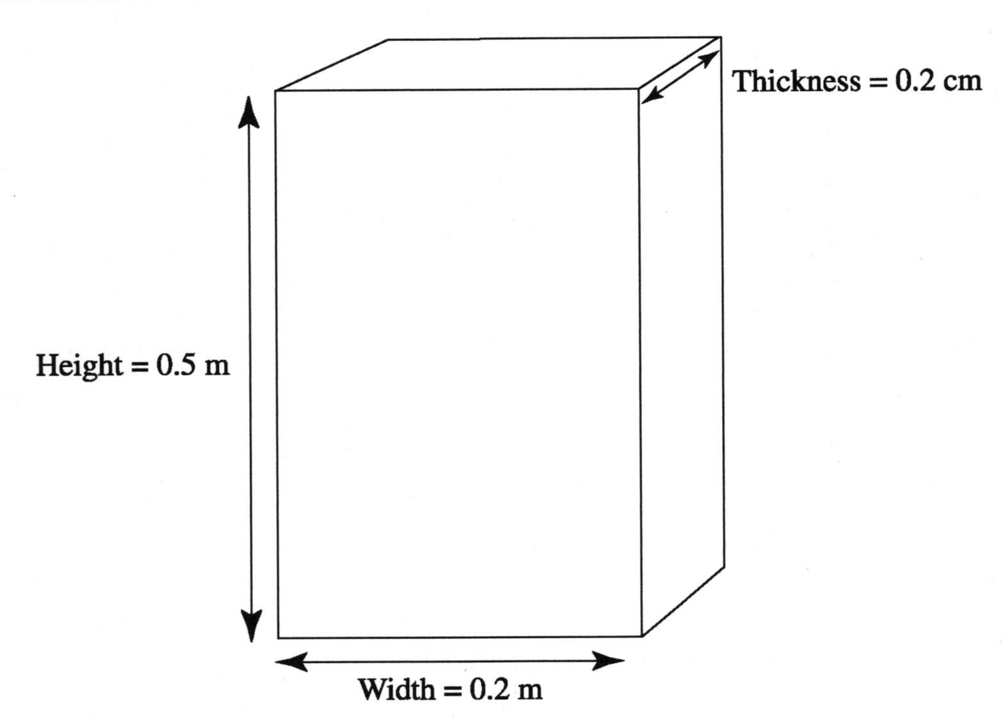

   a. Calculate the force due to water pressure on the top of the block.

   b. Calculate the force due to water pressure on the bottom of the block.

Name ______________________________ Date ____________ Class ____________

c. You can safely say that the horizontal forces due to water pressure on the block sum to zero. Explain why this is.

d. Calculate the net force resulting from water pressure only experienced by the block.

e. Calculate the volume of water displaced by the block.

f. Find the buoyant force on the block.

g. Why would you expect your result from the previous problem match your answer from part d?

h. Predict how the buoyant force would change if the block were rotated flat.

i. Calculate the force due to water pressure on the top of the block.

Name ______________________ Date ____________ Class ____________

j. Calculate the force due to water pressure on the bottom of the block.

k. Calculate the net force resulting from water pressure only experienced by the block.

l. Was your prediction correct? How did this answer compare to your previous results?

5. The flasks shown below are all filled with water. Calculate the pressure on the bottom of each flask.

h = 10 cm
r = 40 cm

h = 20 cm
r = 20 cm

h = 40 cm
r = 40 cm

h_1 = 25 cm

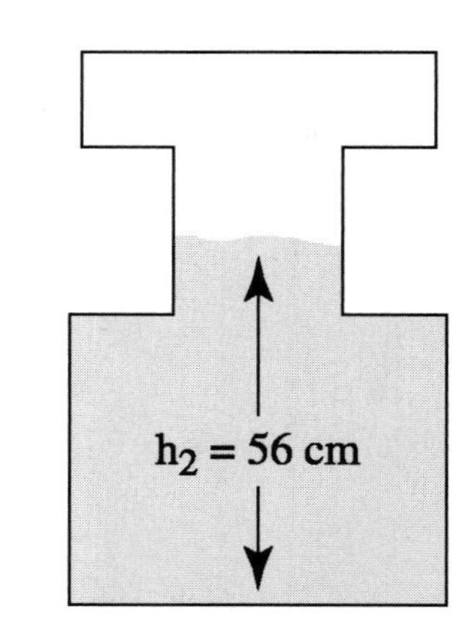

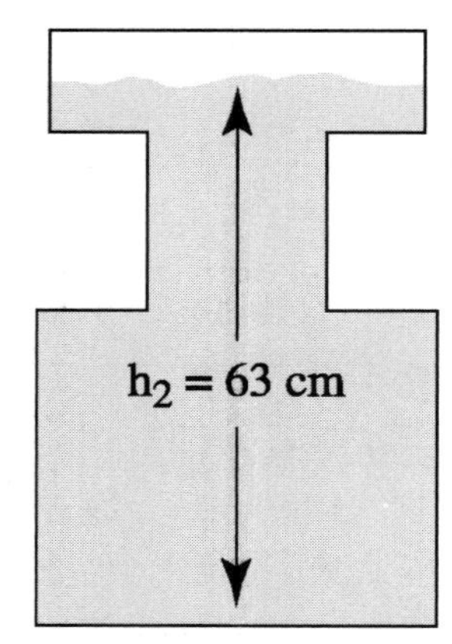

Name ______________________________ Date ____________ Class ____________

6. A cup with a solid metal cube inside is floating in a beaker filled with water. The beaker is sitting on a scale.

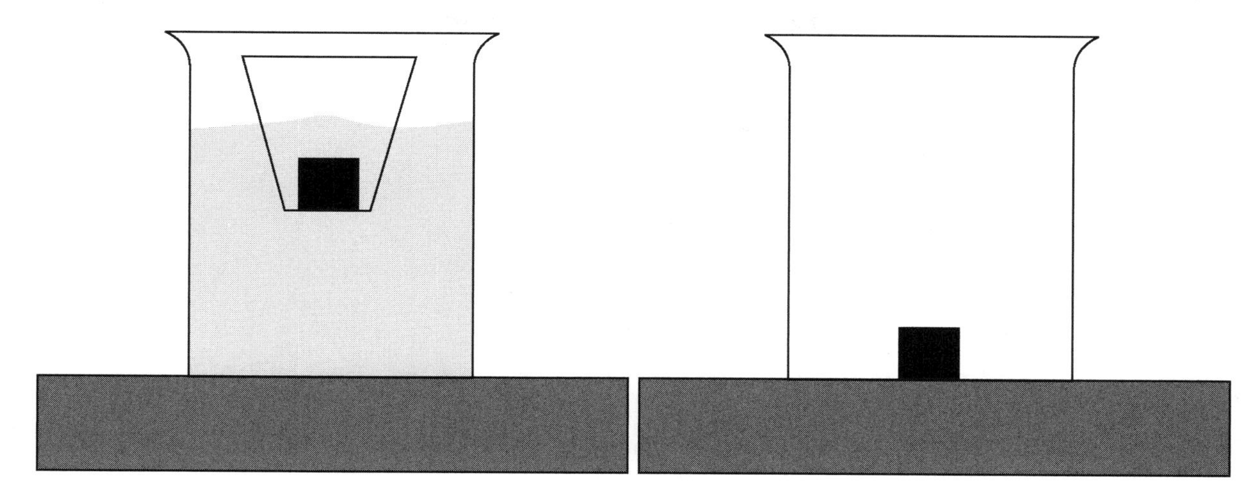

a. Does the reading on the scale change if the cube is removed from the cup and then dropped in to the water? Explain your reasons for choosing your answer.

b. Does the water level in the beaker change when the cube is removed from the cup and dropped in the water? Explain your reasons for choosing your answer and draw the water level in the beaker. Also draw the location of the cup.

c. The cup is pushed down so the lip of the cup is just above the surface of the water. Does either the reading on the scale or the level of the water change? Explain your reasons for choosing your answer.

d. The fully submerged cup springs a leak and fills with water, eventually coming to rest on the bottom of the beaker. Does either the reading on the scale or the level of the water change? Explain your reasons for choosing your answer.

Name ______________________ Date ____________ Class ____________

e. Compare the scale reading and water level with the reading and level when the cube was in the cup and the cup was still floating. Explain.

7. Each of the following containers is filled with the same amount of water. The containers are all open on top and have a small hole near the bottom. Rank, from slowest to fastest, how quickly each container empties completely.

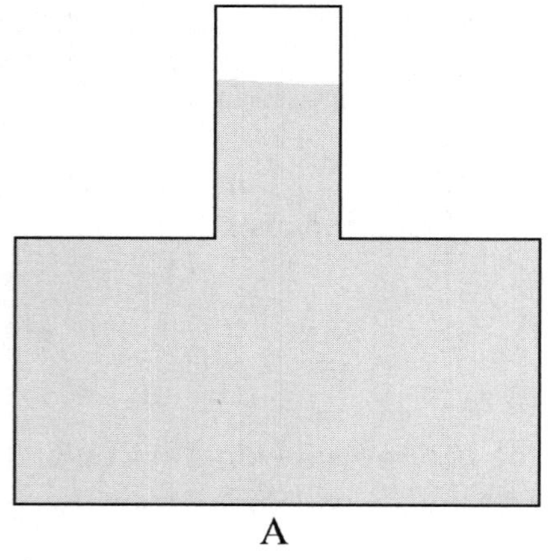

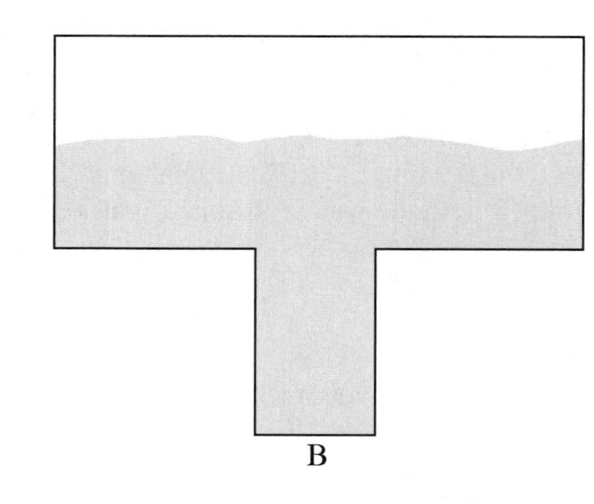

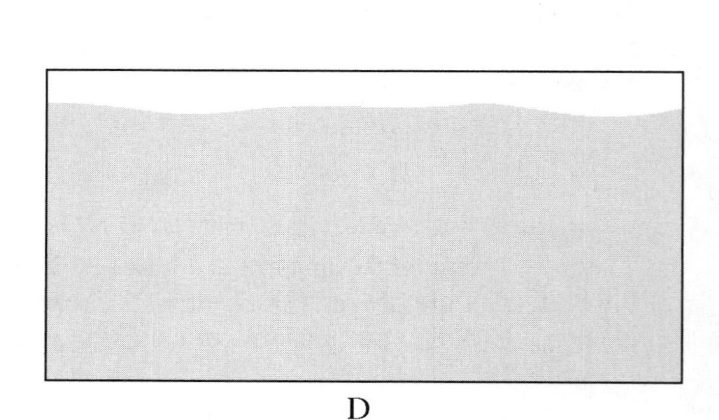

Name ______________________________ Date ____________ Class ____________

a. Explain your choice of ranking in terms of pressure differences.

8. A bottle with two holes in its side is filled with water.

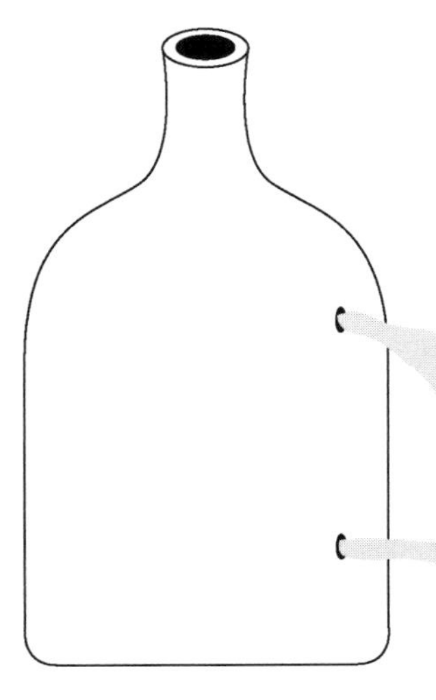

a. Water exiting through the top hole does not spray as far as water coming out of the bottom hole. Explain in terms of pressure differences why this happens.

b. Explain this behavior in terms of energy.

c. Explain why it is easier to figure out the water flow rate out of each hole using energy calculations rather than the forces exerted on the water. Assume you know the dimensions of the bottle, the holes, and the amount of water in the bottle.

Name ______________________________ Date ____________ Class ____________

9. A three lane highway suddenly narrows down to one lane for a short bit before opening up into two lanes. Three friends decide to count cars to pass the time.

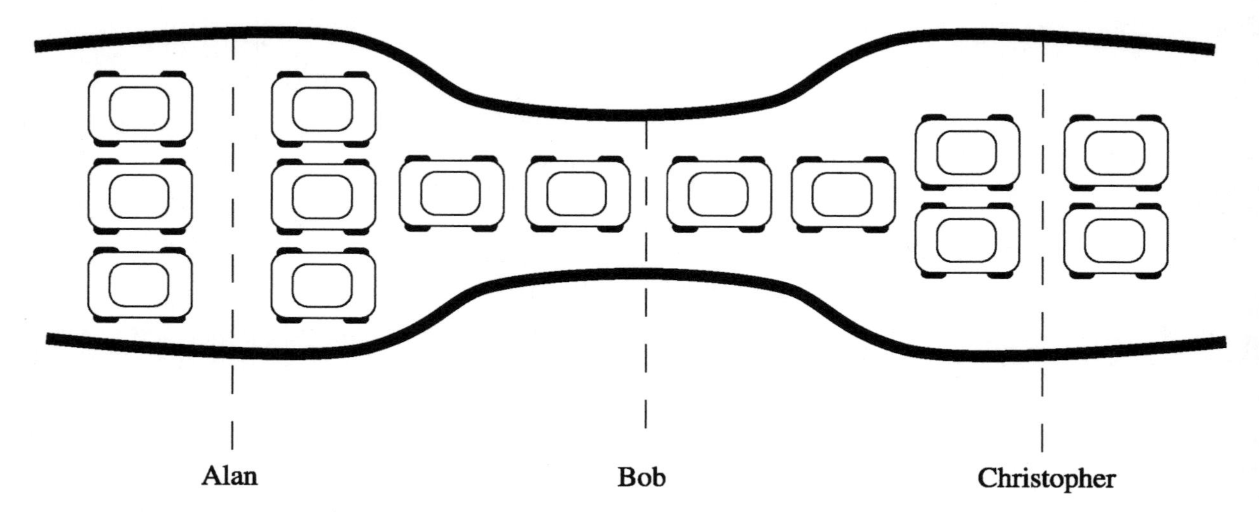

a. Alan says to Bob, "You should count only 1/3 of the number of cars I count since I have three lanes of traffic, while you only have one lane to count. Christopher should count 2/3 of the number I see." What is wrong with Alan's argument?

b. Traffic is moving fairly smoothly, and during a ten minute period Alan counts 270 cars going past him. How many cars will Bob and Christopher count during the same time?

c. The cars must go by Bob faster than by Alan if Alan and Bob count the same number of cars. Explain why this is true.

d. How would you expect the speed of the cars going past Christopher to compare to the speeds Bob and Alan observe? Explain.

e. If there is an off-ramp for the cars between Alan and Bob, would you expect Bob to count more cars, fewer cars, or the same number of cars as Alan counts? Explain your reasons.

f. Would you still expect Bob and Christopher to count the same number of cars? Explain.

g. If Christopher consistently counts more cars than Bob, what possible explanation is there for this difference?

h. Explain how the traffic analogy explains fluid behavior in pipes of different diameters.

Name ______________________________ Date __________ Class __________

10. The diagram below is a *top-down* view of a pipe with fluid flowing from left to right. The flow is steady with no turbulence at all.

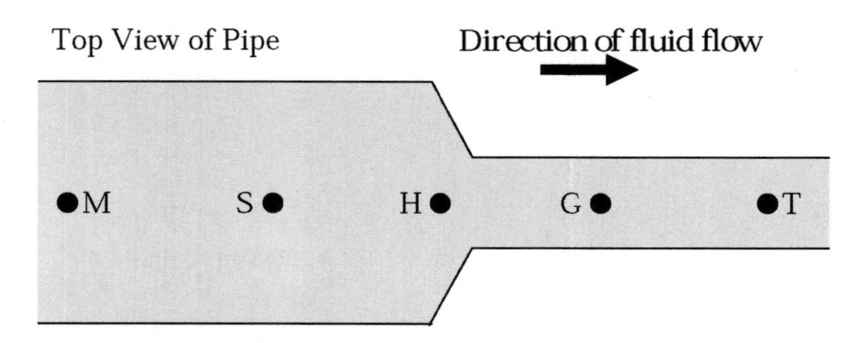

a. Rank the fluid velocity from smallest to largest. Use an equal sign between points with the same velocity. Explain your reasons for choosing your rankings.

b. Rank the fluid pressure from smallest to largest. Use an equal sign between points with the same pressure. Explain your reasons for choosing your rankings.

Name ______________________________ **Date** ___________ **Class** ___________

11. The diagram below is a *top-down* view of a pipe with fluid flowing from left to right. The flow is steady with no turbulence at all.

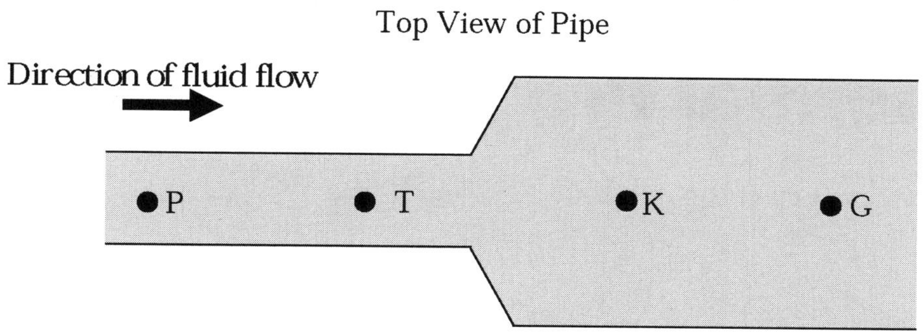

Top View of Pipe

Direction of fluid flow

P T K G

a. Rank the fluid velocity from smallest to largest. Use an equal sign between points with the same velocity. Explain your reasons for choosing your rankings.

b. Rank the fluid pressure from smallest to largest. Use an equal sign between points with the same pressure. Explain your reasons for choosing your rankings.

Name ______________________ Date ____________ Class ____________

Chapter 11

Thermal Properties of Matter

1. Calculate the temperature change between 33°C and 97°C.

2. Convert 33°C and 97°C to Kelvin.

3. Calculate the temperature change between the two temperatures you calculated in the previous problem.

4. Why is the temperature change the same numerical value in Kelvin and Celsius?

5. Several blocks of the same unknown metal but different sizes undergo temperature changes. Rank them from the largest amount of heat transfer to the least amount needed for the temperature change.

   Block A: 11°C to 38°C, mass m
   Block B: 25°C to 41°C, mass m
   Block C: 123°C to 131°C, mass 1.5m
   Block D: -23°C to 11°C, mass 0.75m
   Block E: 97°C to 103°C, mass 1.75m

6. Several blocks of different mass and material all undergo the same temperature change. Rank them from the largest amount of heat transfer to the least amount needed for the temperature change. All masses are given relative to the mass of the iron block.

   Block A: Iron block with mass m
   Block B: Copper with a mass of 2m
   Block C: Gold with a mass of 1.5m
   Block D: Aluminum with a mass of 0.7m
   Block E: Tungsten with a mass of 1.5m

Name ______________________________ Date ____________ Class ____________

7. A beaker of room temperature water is placed on a Bunsen burner and eventually the water starts boiling. A coil of copper wire at room temperature is dropped into the water and allowed to sit for a few minutes. The copper wire is then removed and quickly dropped into a calorimeter filled with cold water.

   a. The water is not in direct contact with the flame, so what is the source of the heat transferred to the water?

   b. Is the beaker in thermal equilibrium with the burner flame? Explain your reasoning for your answer.

   c. Is the water in thermal equilibrium with the beaker? Explain your reasoning for your answer.

   d. When the copper coil is dropped in the water, list all of the heat transfers that occur (indicate which heat transfers are energy losses (-Q) or energy gains (+Q)).

   e. The copper wire is in thermal equilibrium with the boiling water before it is removed. List all the heat transfers as the coil is removed from the boiling water and dropped into the calorimeter.

8. Should a fluid used as a coolant in an engine have a large or small specific heat? Explain how specific heat affects the cooling rate.

Name ______________________________ Date ____________ Class ____________

9. Two blocks of different materials (but the same mass) are dipped into equivalent beakers of water. The final water temperature in beaker A is larger than the final temperature in beaker B. Did block A or block B have a larger specific heat? Explain how you came to your answer.

10. At night, the ocean cools off faster than the land, frequently causing a breeze. Can this temperature difference be explained by the difference in specific heat of water and the ground? Explain your reasoning.

11. A piece of copper is heated to just below its melting temperature and then dropped in a beaker of room temperature water. Explain why each of the following statements is true or false.

    a. Since the melting point of copper is so much higher than the boiling point of water, the water will come to a boil fairly quickly.

    b. The specific heat of copper is much smaller than the specific heat of water, so the water will not come to a boil.

12. Does work need to be done on water for a change of state to occur?

Name ______________________ Date ____________ Class ____________

Chapter 12

The Kinetic Theory of Gases, Entropy, and Thermodynamics

1. In the previous chapter, calculations involving changes in temperature yielded the same result for Kelvin and Celsius. Does it matter whether you use Kelvin or Celsius with the ideal gas law equation? Give a numerical calculation to support your answer.

2. Canned air is frequently used to blow dust off of computers or other delicate electronic circuits.

   a. If a long burst of air is expelled, the outside of the can gets fairly cold. Explain in terms of the molecules inside the can why the outside might get cold.

   b. In terms of the ideal gas law, give a possible explanation of why the outside of the can gets cold.

3. Are over-inflated tires more likely to burst on hot days or cool days? Explain your reasons.

4. A cylinder sealed by a piston is filled with an ideal gas. The piston can freely slide inside the cylinder and is of negligible mass. The piston is held in place by a latch as a burner is used to heat up the cylinder and the gas inside it. What happens to the piston when the latch is released?

Name ______________________________ Date ____________ Class ____________

5. Looking at the ideal gas law equation, what changes does a bike tire pump make to the tire to increase the air pressure in the tire?

6. Give three examples of isothermal processes (e.g. using a hand pump to slowly pump up a bike tire).

7. Give three examples of an adiabatic process (e.g. letting out a burst of air from a can of air).

8. Can you have a process that is both isothermal and adiabatic? Explain your reasons.

Name ______________________________ Date ____________ Class ____________

Chapter 13

Periodic Motion and Simple Harmonic Oscillators

1. A quick review: Each case below depicts a block moving on a frictionless surface. Sketch a possible x versus t graph for each case. The pointer is at $x = 0$ m and x-values to the right of that point are positive. Assume constant velocity if no acceleration vector is shown.

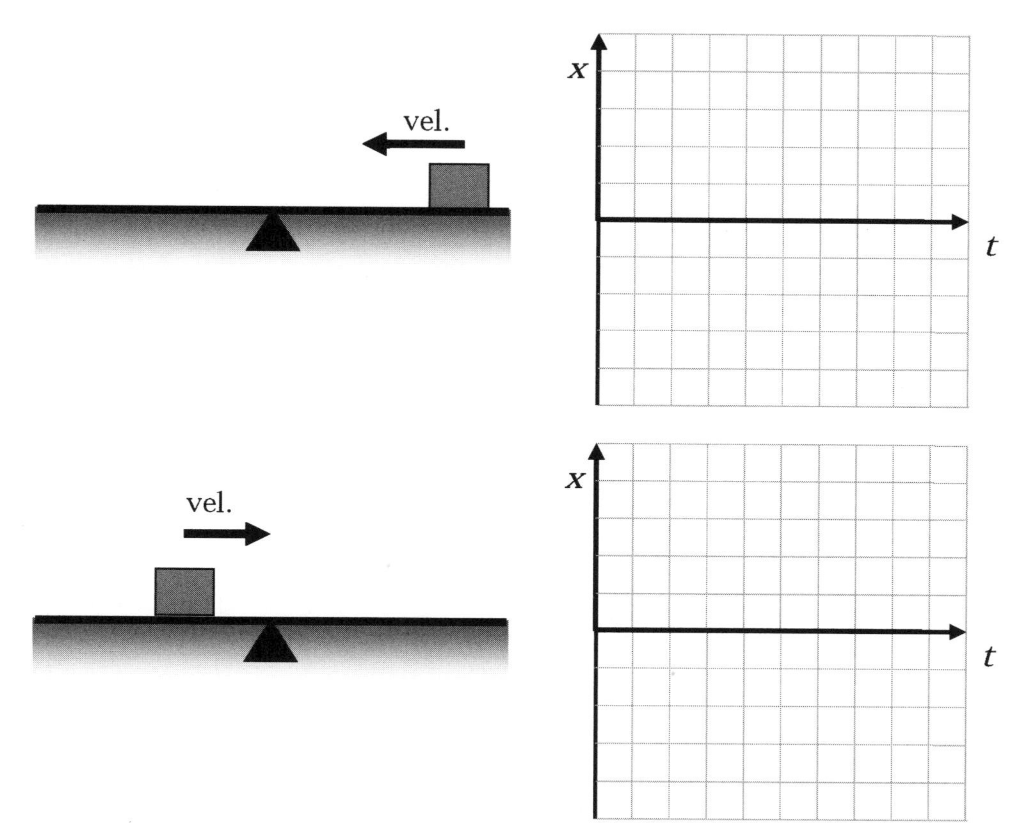

Name ______________________________ **Date** ____________ **Class** ____________

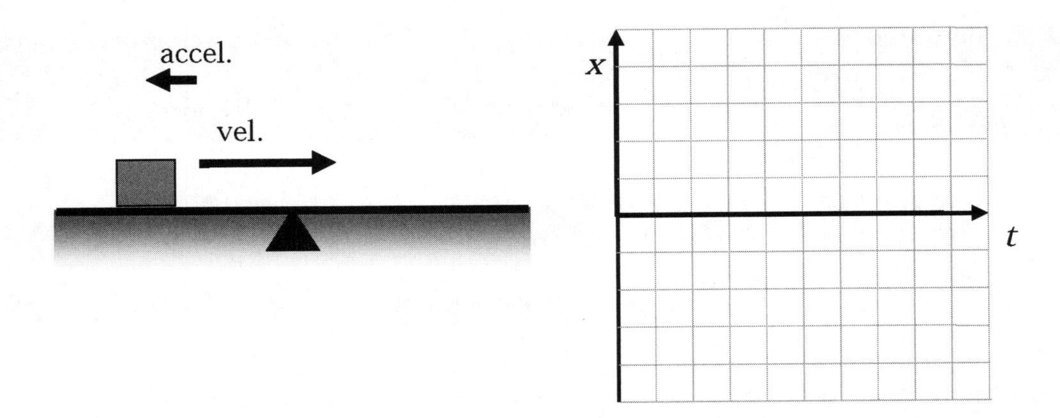

2. The diagram below shows a ball following a circular path. The number next to each ball represents the clock time in seconds.

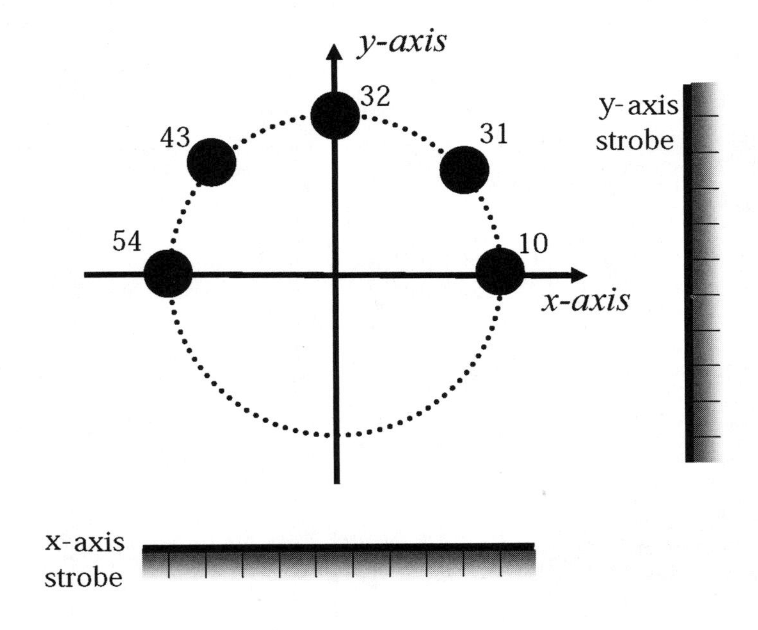

a. How long does it take for the ball to complete one full revolution?

b. Use your answer to the previous question to calculate how many revolutions the ball makes each second. Your answer will be less than one. What does it mean to have a fractional answer?

Name ______________________________ Date ____________ Class ____________

c. Draw the strobe picture for the ball during the times shown on the x-axis and y-axis on the previous diagram.

d. Use the strobe pictures to draw x versus t on the graph below. Use different colors to indicate where the ball is moving fastest, where it is moving slowest, and where it changes direction along the x-axis.

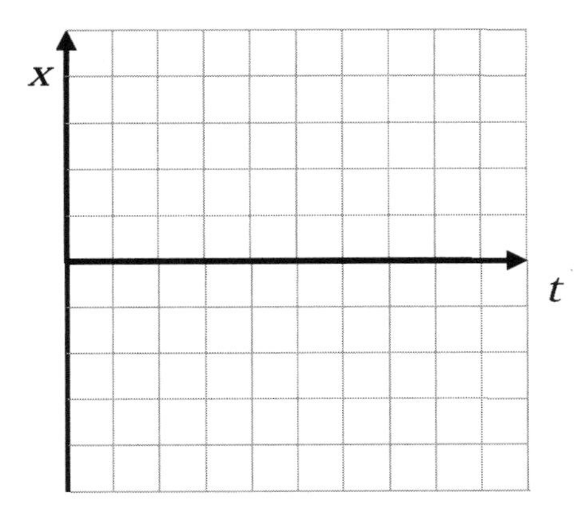

e. Since the ball is moving in a circle, there must be a radial force exerted on the ball. Draw the direction of the radial force on the ball at each of the five times in the diagram below.

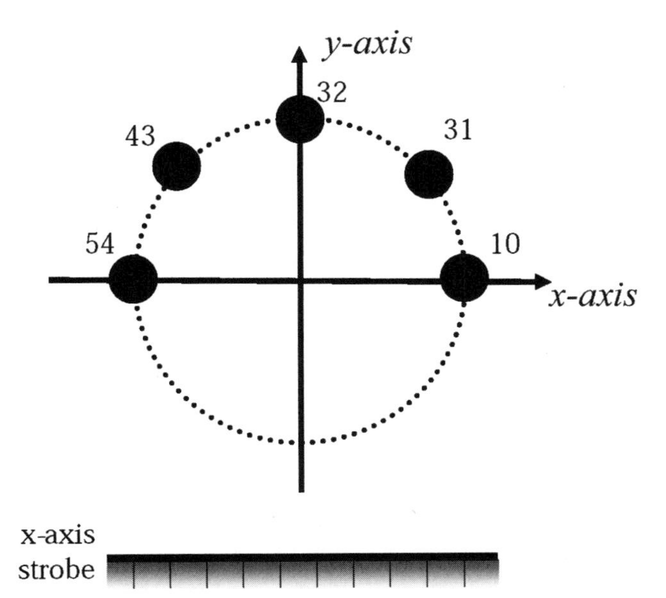

Name ______________________ Date ______________ Class ______________

f. On the x-axis strobe picture, draw a vector for the component of the radial force along the x-axis for each time.

3. A block is connected to a spring and is free to slide back and forth on frictionless surface.

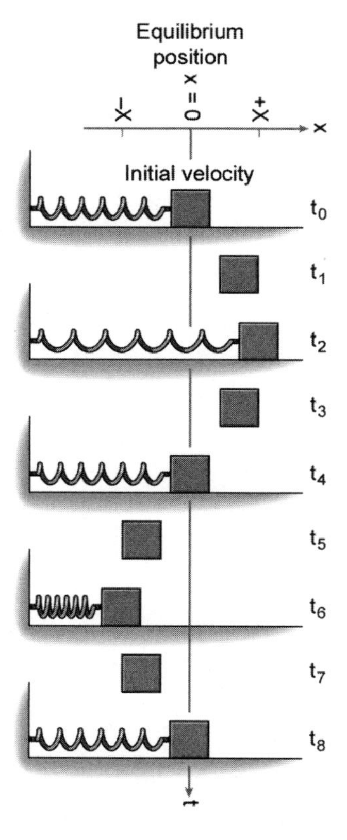

4. Draw the free body diagram for the block at times t_0, t_3, t_4, t_5, and t_7. Use another color to indicate the net force for each time.

5. Why should the free body diagram for t_0 and t_4 look identical?

6. Are the accelerations of the block at t_0 and t_4 the same?

Name ______________________________ **Date** ____________ **Class** ____________

7. Are the velocities of the block at t_0 and t_4 the same?

8. Of the previous free body diagrams, which other pair of diagrams are identical?

9. Are the accelerations the same for those two times?

10. Are the velocities the same for those two times?

11. If the velocity and acceleration of the block are in opposite directions, what is happening to the speed of the block?

12. If the velocity and acceleration of the block are in the same direction, what is happening to the speed of the block?

13. Sketch the displacement y versus t graph for the block. Indicate on the graph where the block is moving fastest and where it is moving slowest.

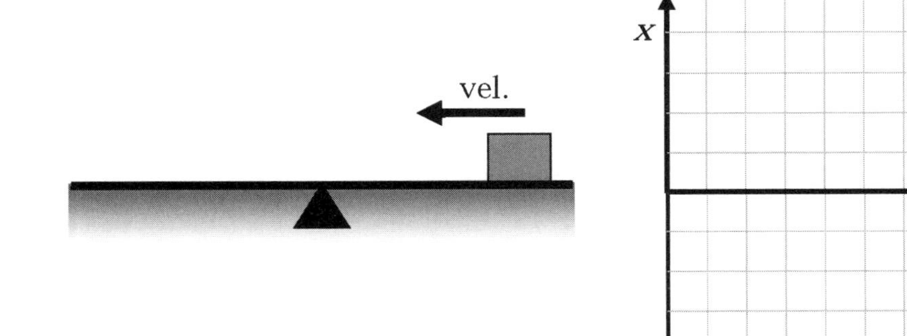

14. Sketch the velocity v versus t based on the slope of the previous graph. Indicate on the graph where the velocity is changing the most and where the velocity is changing the least.

15. Sketch the acceleration a versus t graph based on the previous graph.

16. How would the graph of force F versus t differ from the acceleration graph?

Name ______________________ Date __________ Class __________

Chapter 14

Waves and Sounds

1. Which types of waves (longitudinal, transverse) can be made easily using the materials listed below? Explain how you would make each wave type.

    a. Slinky

    b. Rope

    c. Air in a tube

    d. Water in a tub

2. A friend is about to take a quiz on waves and needs a little help. How would you explain the similarities and differences between longitudinal and transverse waves?

Name ______________________ Date ____________ Class ____________

3. Two snap shots of a slinky show the coils before a wave pulse passes and as the pulse is passing, moving to the right. Sketch the displacement *d* of the links along the length of slinky shown here. (See Figure 14-5 in the text for help.)

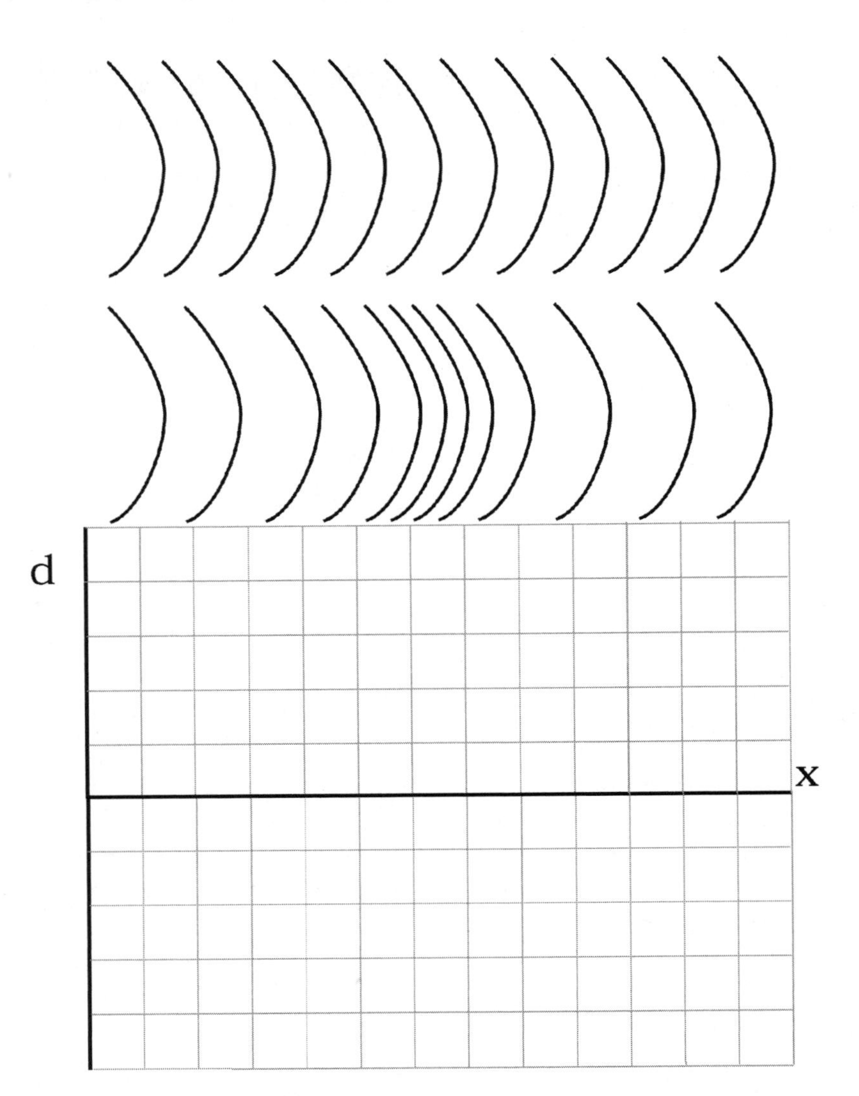

Name ______________________________ Date ____________ Class ____________

a. Sketch the graph of the displacement of a single link over time as the pulse moves along the slinky.

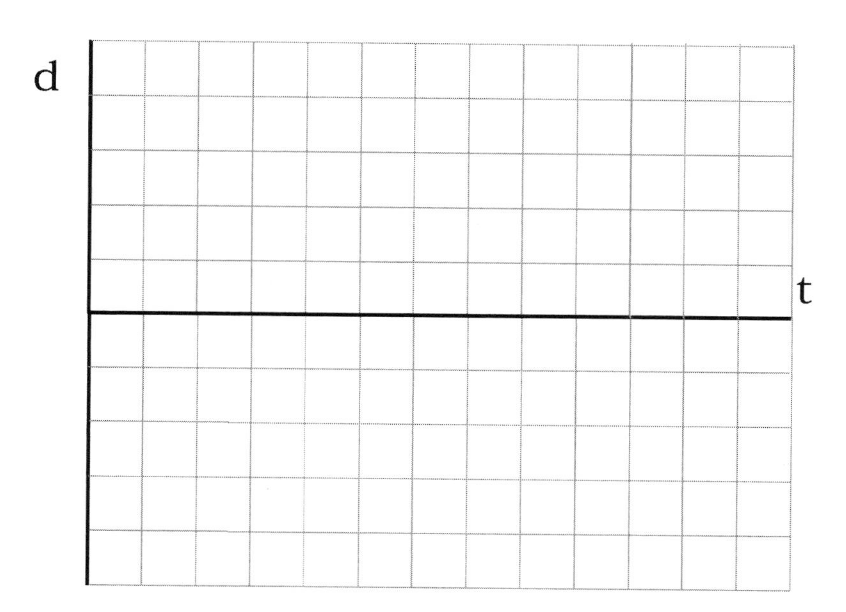

b. Why does this graph look similar to the graph of displacement versus position?

c. What is the difference between propagation speed and displacement speed?

d. In previous chapters you learned that the slope of a position versus clock reading graph is the velocity. Is the slope of the displacement versus clock reading graph the propagation velocity or displacement velocity? Explain your reasons.

Name ______________________ Date __________ Class __________

e. Sketch the velocity graph for the displacement graph you drew previously.

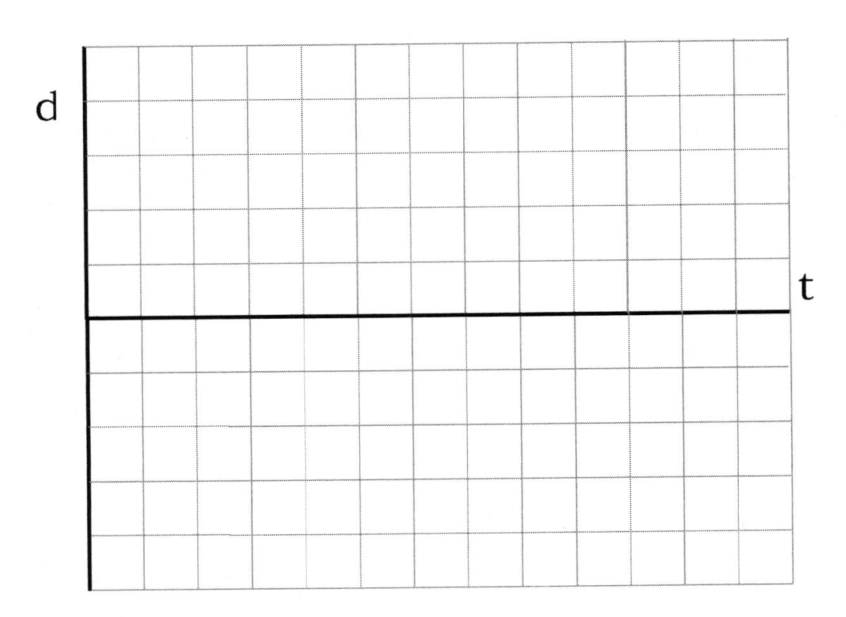

4. Draw the sum of the two wave pulses. (Square and triangular pulses are used in this exercise only because they are easier to work with than other shaped pulses.)

   a. Draw the sum of the two wave pulses.

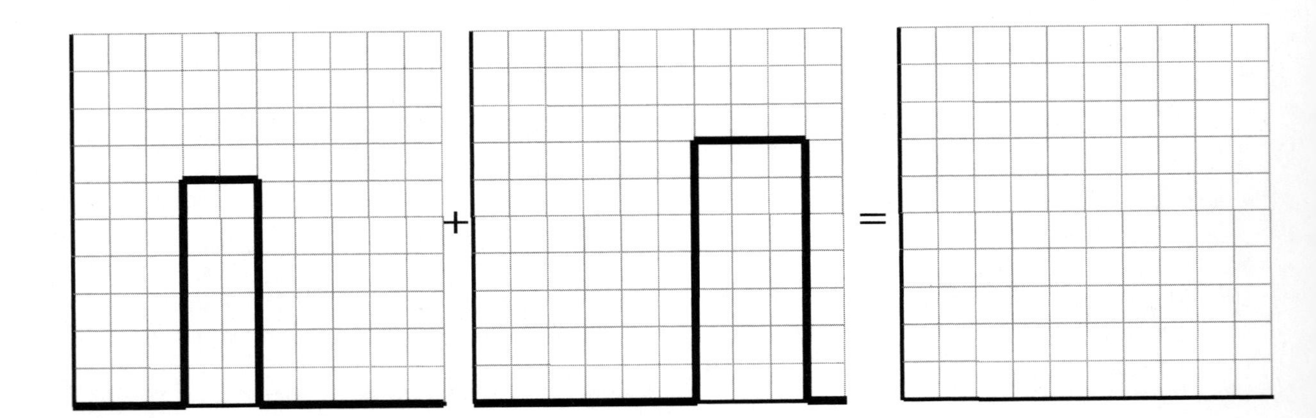

Name ______________________ Date __________ Class __________

b. Draw the sum of the two wave pulses.

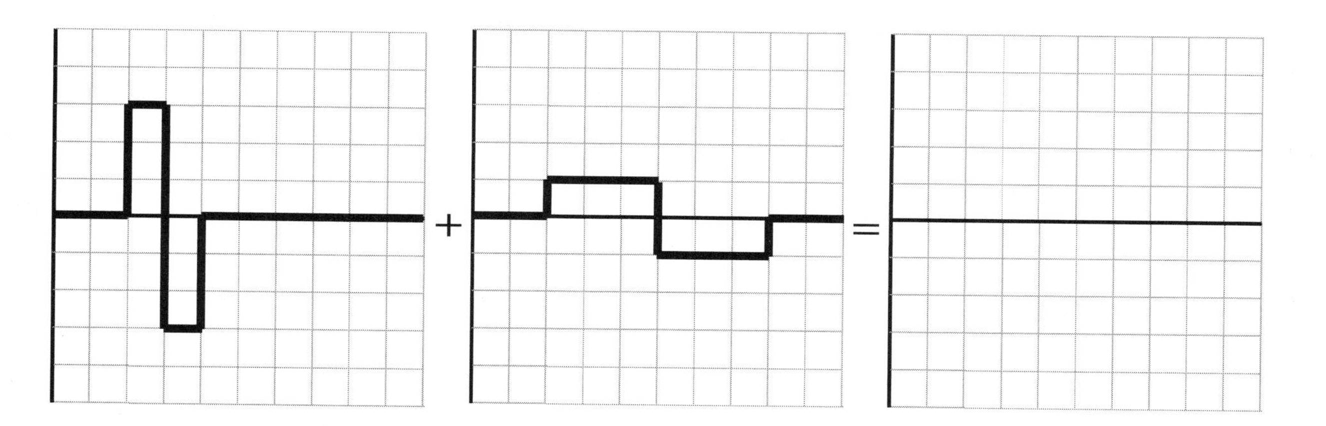

c. Draw the sum of the two wave pulses.

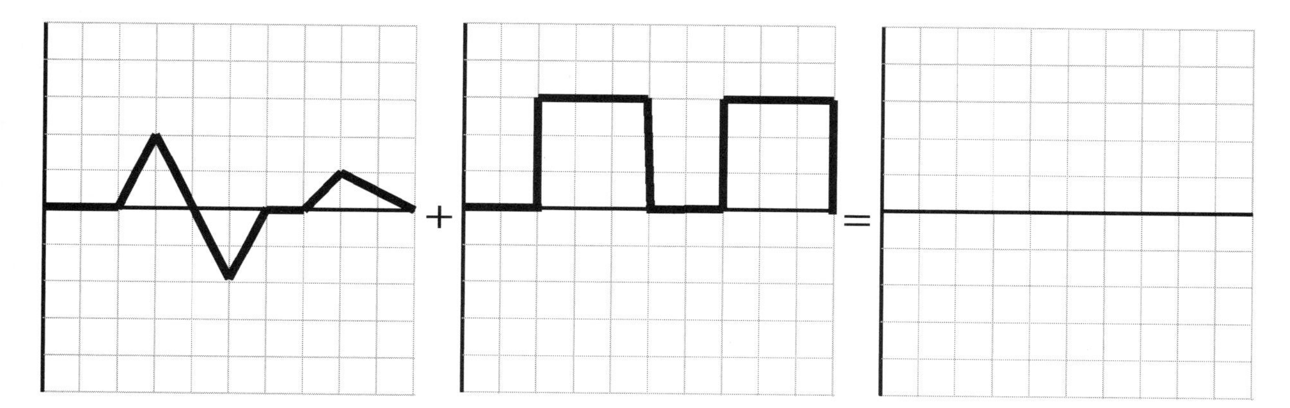

Name ______________________ Date ____________ Class ____________

5. Draw the sum of the two waves. Assume the wave trains are very long.

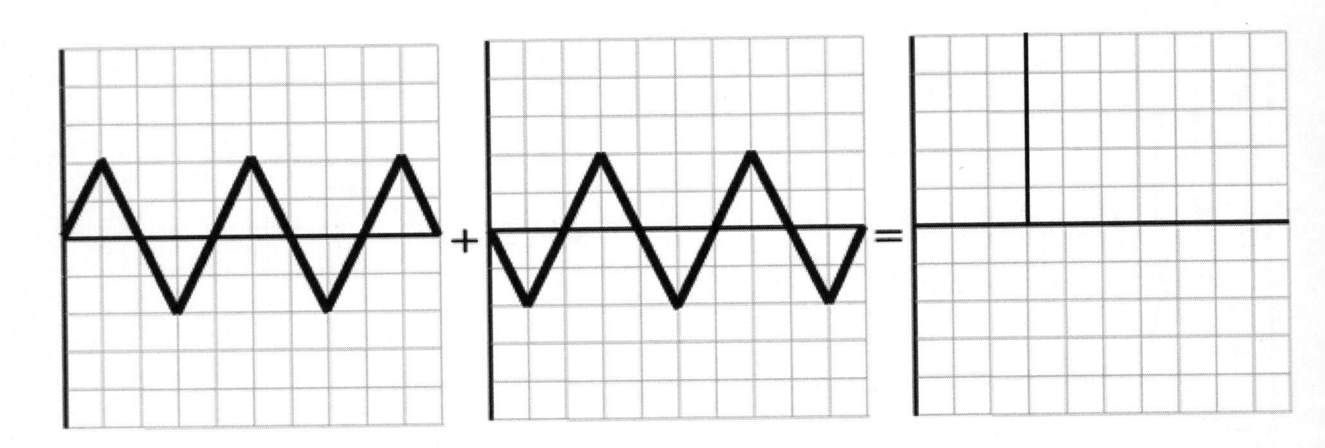

a. The two waves are identical except the right hand wave is shifted. How many squares is the right hand wave advanced from the left hand wave? What fraction of a wavelength is this?

b. Sketch a wave that is advanced by 3 squares and then add the two waves together. What fraction of a wavelength is 3 squares?

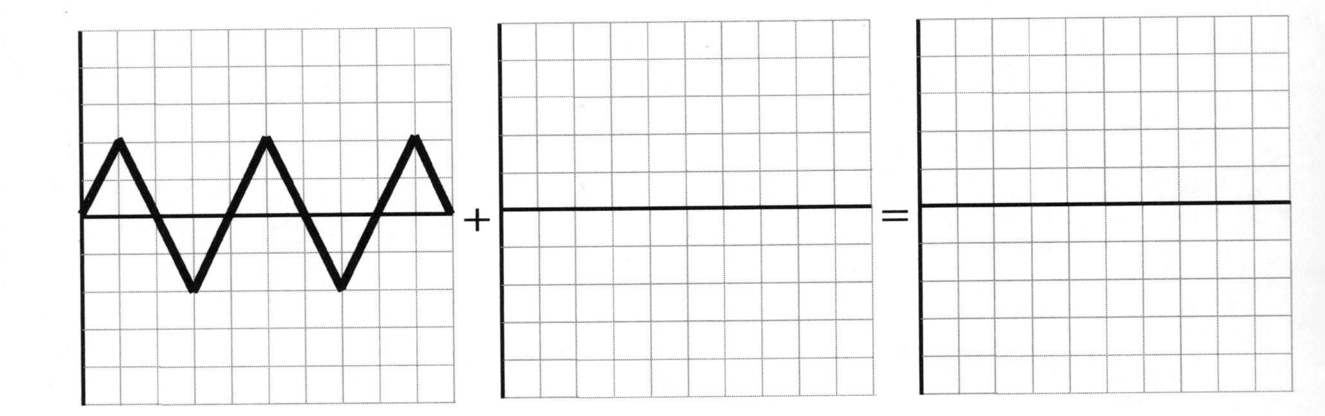

Name ______________________________ Date ____________ Class ____________

c. Sketch a wave that is advanced by 4 squares and then add the two waves together. What fraction of a wavelength is 4 squares?

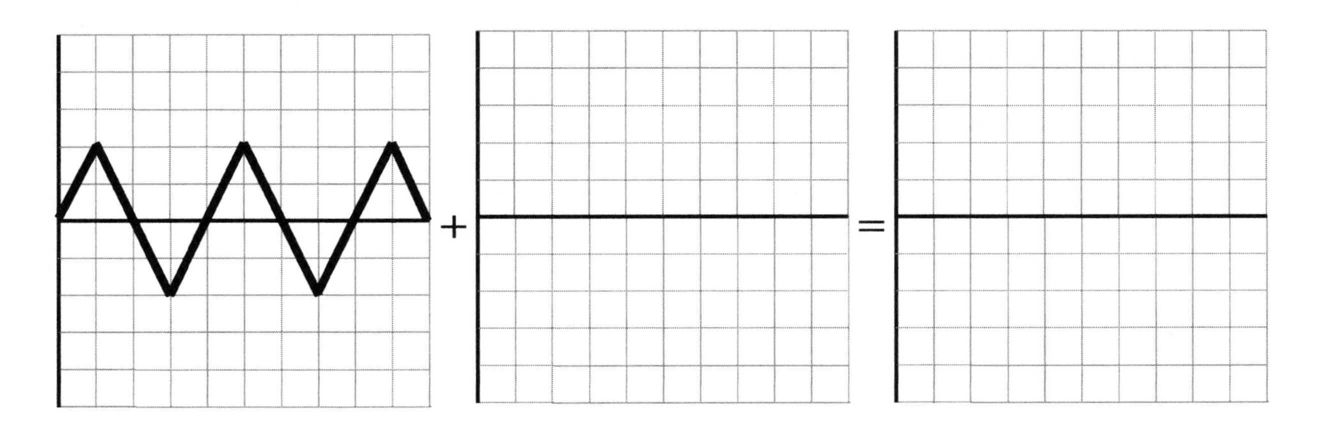

6. Draw the sum of the two sinusoidal waves. Add the amplitudes on the grid lines and sketch the waveform between points.

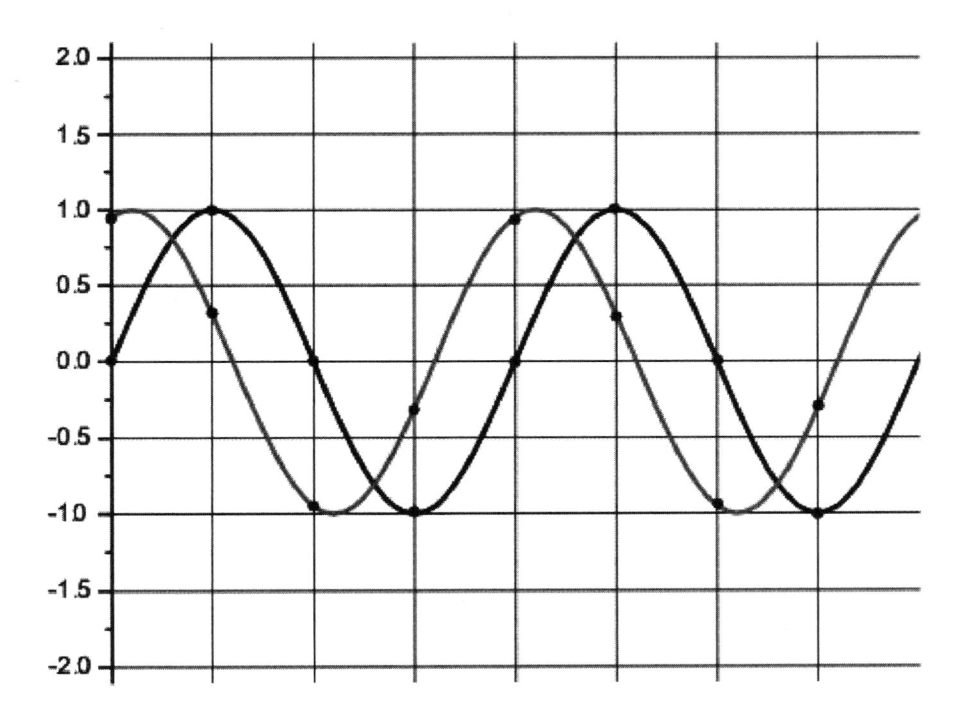

Name ______________________________ Date ____________ Class ____________

7. Draw the sum of the two sinusoidal waves. Add the amplitudes on the grid lines and sketch the waveform between points.

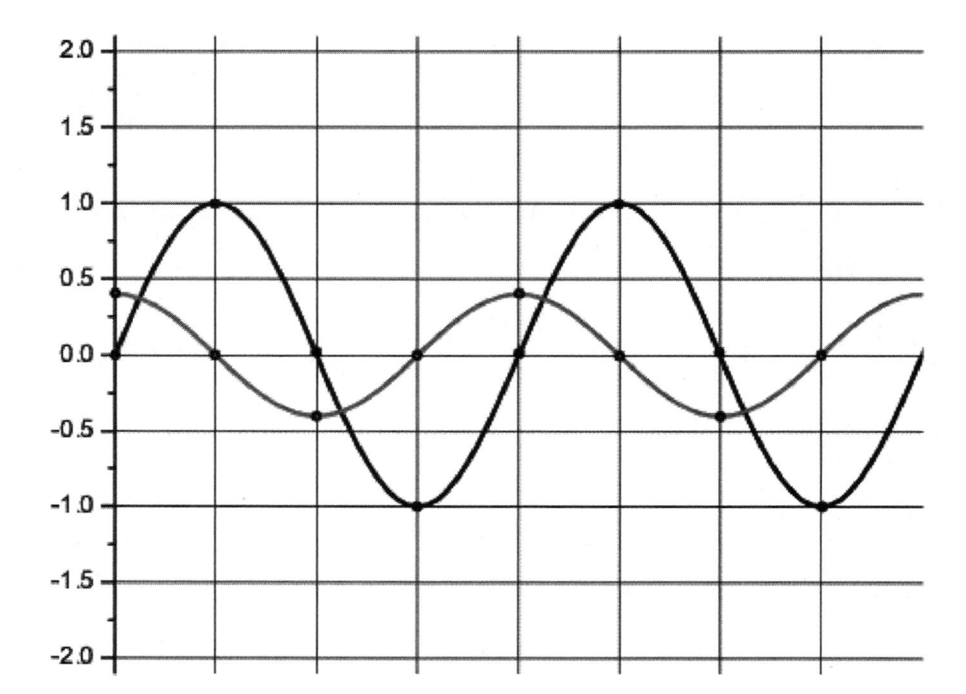

Name ______________________________ Date ____________ Class ____________

8. Sketch the displacement amplitude for the wave on the slinky. The waves move left to right.

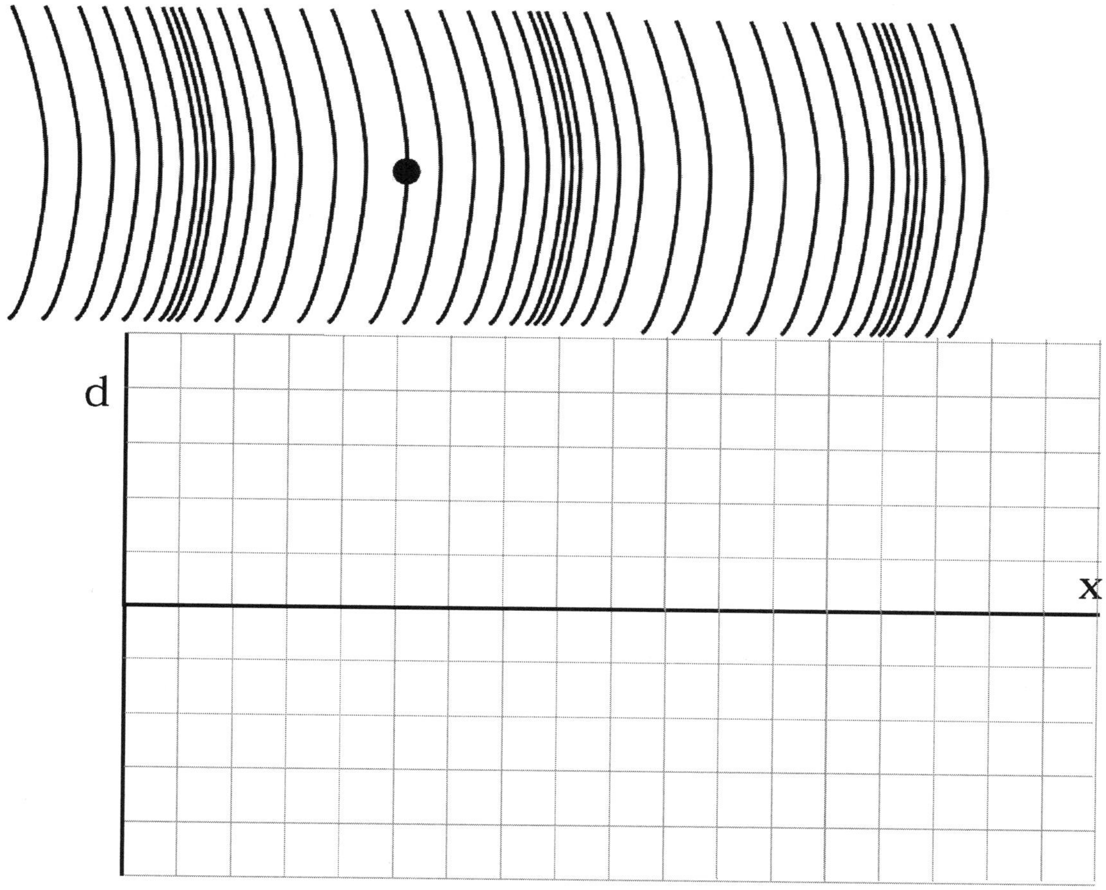

d

x

Name ______________________ **Date** ____________ **Class** ____________

a. Sketch the displacement of the link indicated by the black dot versus time.

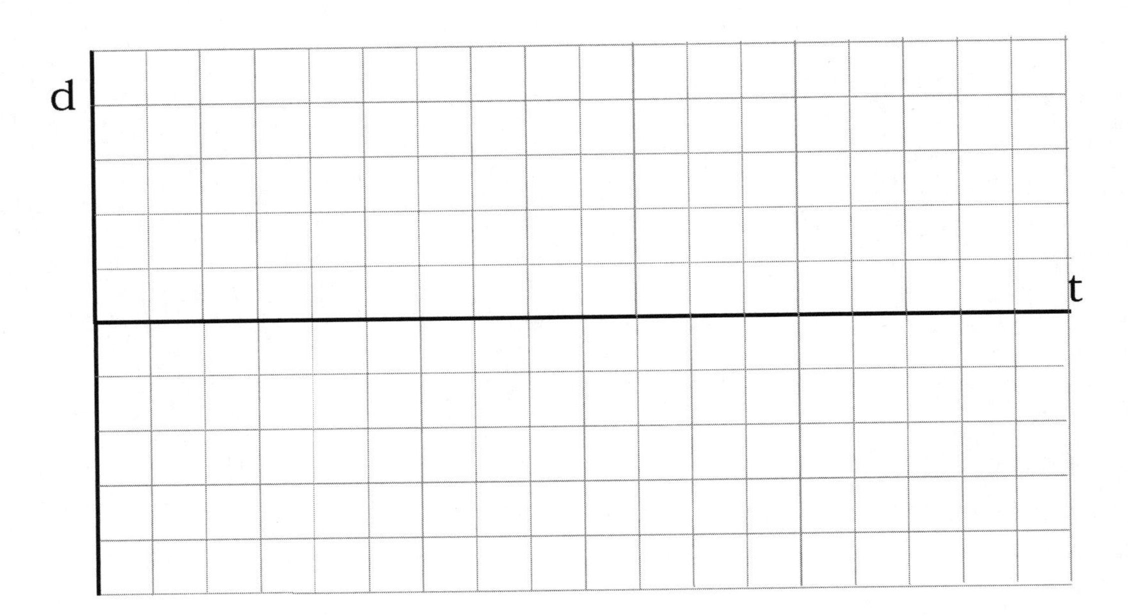

b. Based on your sketch of the displacement versus time graph, draw the velocity versus time graph. Label whether this is the displacement velocity or propagation velocity.

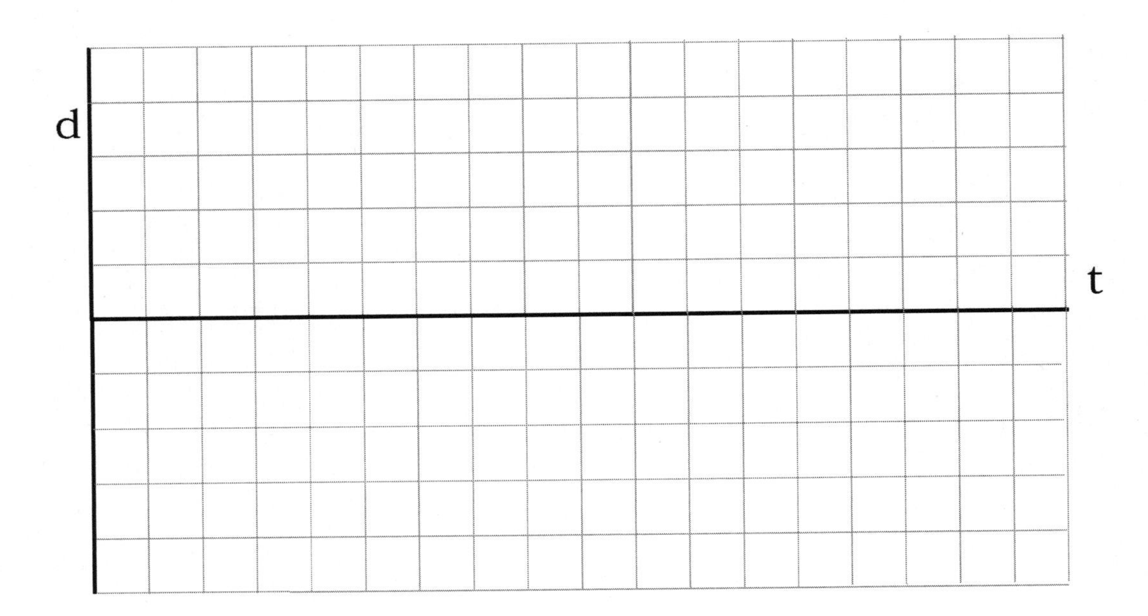

Name ______________________________ Date ____________ Class ____________

c. How is the v versus t graph related to the acceleration a versus t graph?

d. Use the previous v versus t graph to sketch the a versus t graph. What is being accelerated?

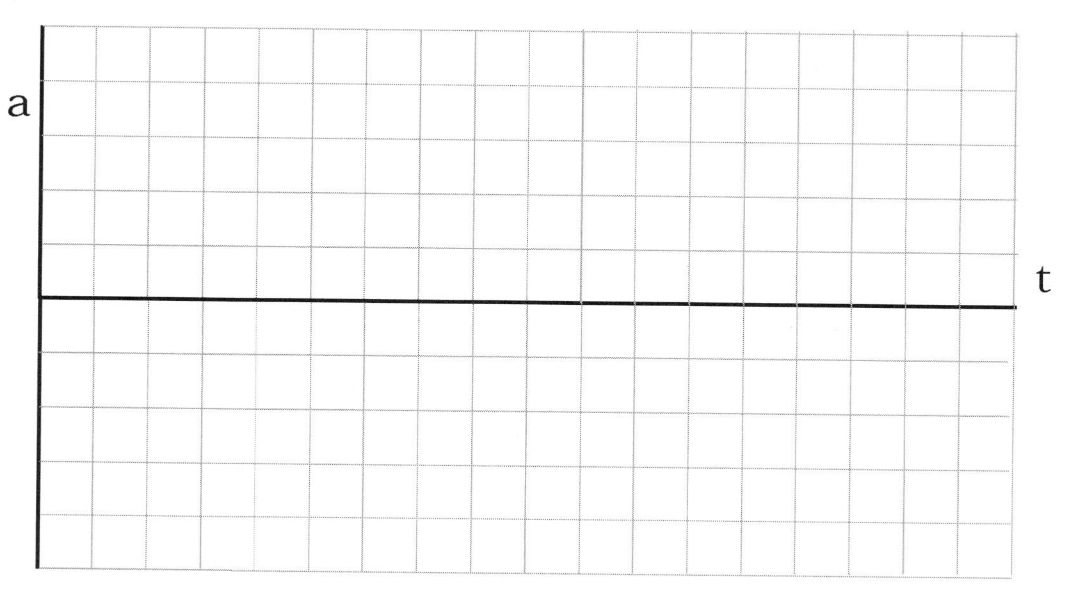

e. How would a graph of force F versus t look different from the a versus t graph? Explain your answer.

9. A single pulse travels from left to right on the slinky shown on the following page.

a. When is the velocity of a single link greatest, when the link displacement is largest or when it is zero? Explain your answer.

b. How is the displacement of a single link connected to the spring potential energy of that link?

c. Fill in the energy bars for kinetic energy, elastic potential energy, and total energy of the slinky ring at each of the three points.

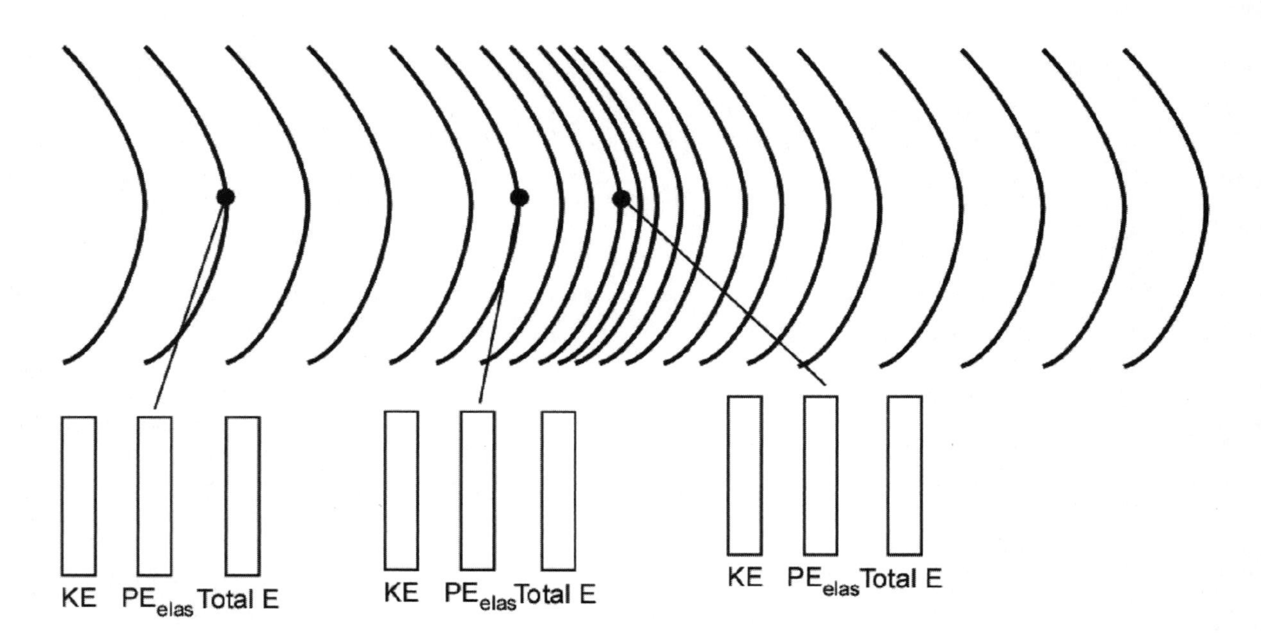

d. Compare the total energy of a link after the pulse has passed to the total energy at the peak of the pulse. The total energy for a single link is not conserved. What system would you need to consider for the total mechanical energy to be conserved?

10. The equation for calculating displacement from a sinusoidal wave function is:

$$y = Y \sin 2\pi \left(\frac{t}{T} - \frac{x}{\lambda} \right)$$

a. Identify which of the letters are variables and which are constants that depend on the physical wave.

Name ______________________ Date ____________ Class ____________

11. A transverse wave travels along a 3.5 m slinky. The amplitude is 2.0 cm and the wavelength is 11.0 cm. You noted earlier that a single pulse travels the length of the slinky and is reflected back to you in 4 s.

   a. How fast does the wave travel along the slinky?

   b. Calculate the period of the wave.

   c. Write the equation for a sinusoidal wave function using the numbers you just calculated and the values provided.

   d. Plot y versus x for this wave. Label the axes and the units.

   e. For each of the following changes, describe how the wave would change. Include a rough sketch of what the wave would look like.

      i. The period is significantly decreased.

      ii. The wavelength is doubled.

      iii. The amplitude is 10 times smaller.

Name ______________________ Date ____________ Class ____________

Chapter 15

Wave Optics

1. Draw the sum of the two waves. Assume the wave trains are very long.

   a. Sketch a wave that is advanced by 5 squares and then add the two waves together. What fraction of a wavelength is 5 squares?

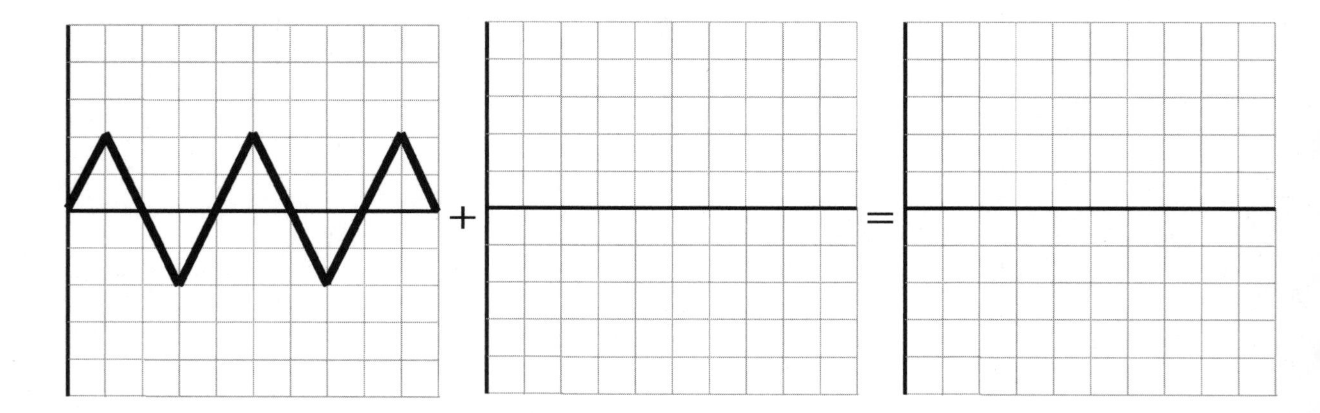

   b. Sketch a wave that is advanced by 21 squares and then add the two waves together. What fraction of a wavelength is 21 squares?

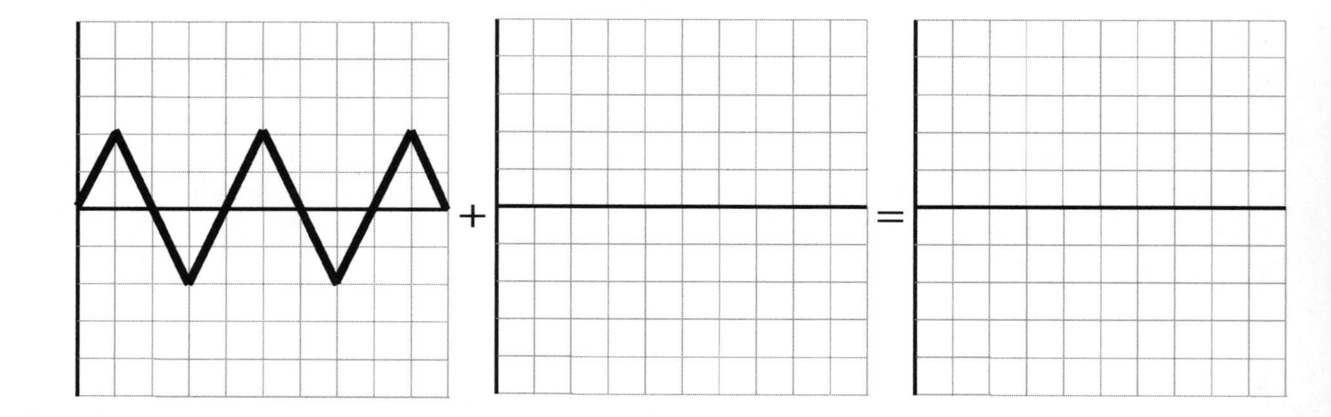

Name ______________________________ Date ____________ Class ____________

c. How many wavelengths correspond to 16 squares?

d. If the second wave was advanced 16 squares, what would the sum look like?

e. How many wavelengths correspond to 10 squares?

f. If the second wave was advanced 10 squares, what would the sum of the two waves look like?

g. Suppose the second wave was shifted by an integral number of wavelengths, say n = 13 wavelengths, from the first wave. Would the shifted wave look any different than the unshifted wave?

h. Would there be any difference between a wave shifted 2 1/2 wavelengths compared to a wave shifted 9 1/2 wavelengths?

i. The sum of the two waves depends on the fractional wavelength shift and not the integral number of wavelengths shifted. Explain why only the fraction of a wavelength matters for adding two waves together.

Name ______________________________ Date ____________ Class ____________

j. Explain in terms of the number of wavelengths shifted how you can tell whether constructive or destructive interference will occur.

2. Two sources emit waves with the same amplitude, frequency, and phase. The wavelength of the waves is 0.30 cm. The distance from source A to an observation point is 21.00 cm and the distance from source B to that observation point is 16.05 cm.

   a. What is the path difference (in cm) of the two sources from the observation point?

   b. How many wavelengths is source A from the observation point?

   c. How many wavelengths is source B from the observation point?

   d. Since the path traveled by each wave is different, the two waves will be shifted with respect to each other. Use your results from the previous two questions to calculate the path difference of the two waves in terms of the wavelength.

   e. When the two waves are superposed at the observation point, will there be constructive interference, destructive interference, or something in between? Explain your reasoning.

Name ______________________________ Date ____________ Class ____________

3. The rings represent a top view of the waves and troughs of a wave. Sketch the cross section along the line. The solid circle lines represent peaks and the dotted circle lines represent valleys. The vertical dotted lines are visual guides to sketch the peaks.

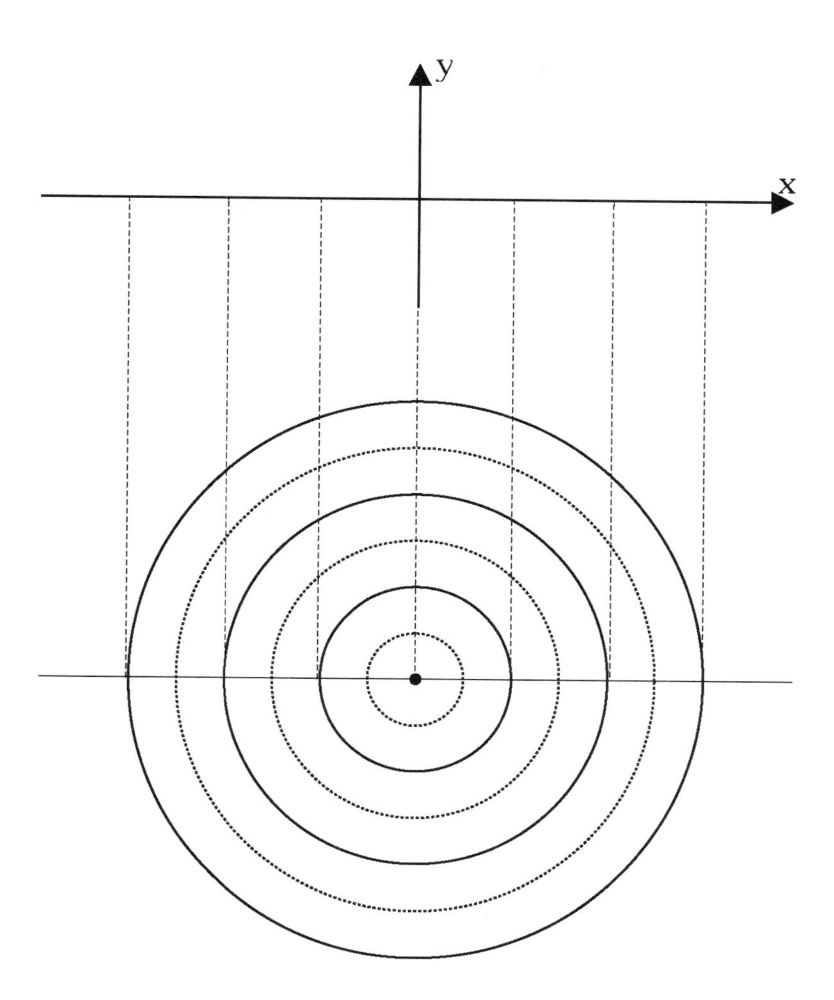

Name ______________________________ Date ____________ Class ____________

4. The diagram below shows the wave crests from two sources. The solid circular lines represent wave peaks and the dotted lines are troughs. Locate where either two peak circles or two trough circles intersect and mark it with a dot. Draw lines along the points to indicate where constructive interference occurs.

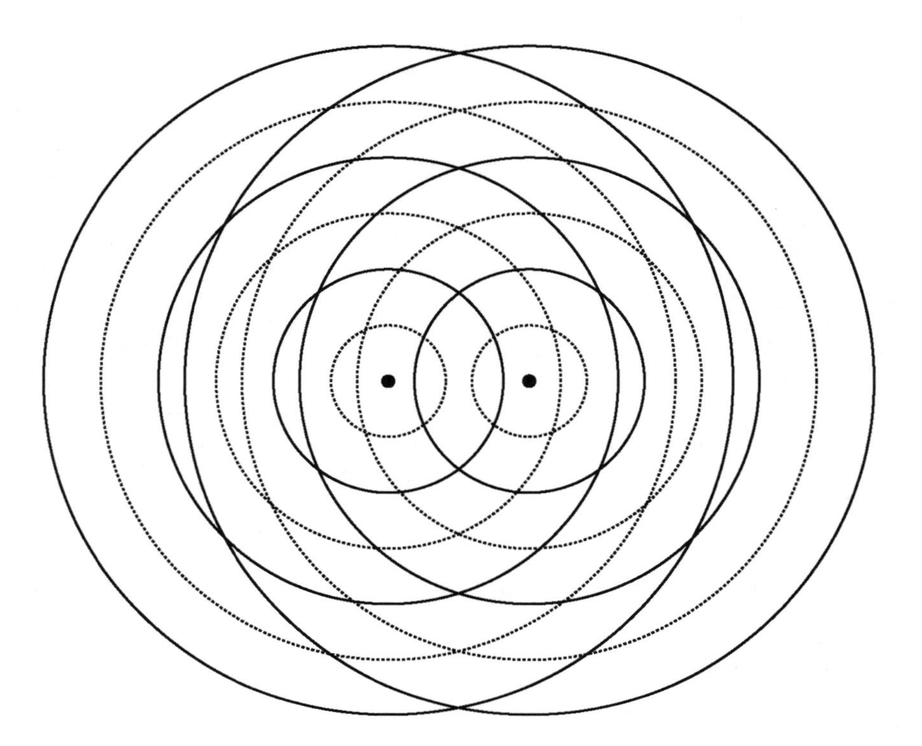

Name ______________________________ Date ____________ Class ____________

a. Mark where each peak and trough intersect and mark it with a dot. Draw lines along the points to indicate where destructive interference occurs.

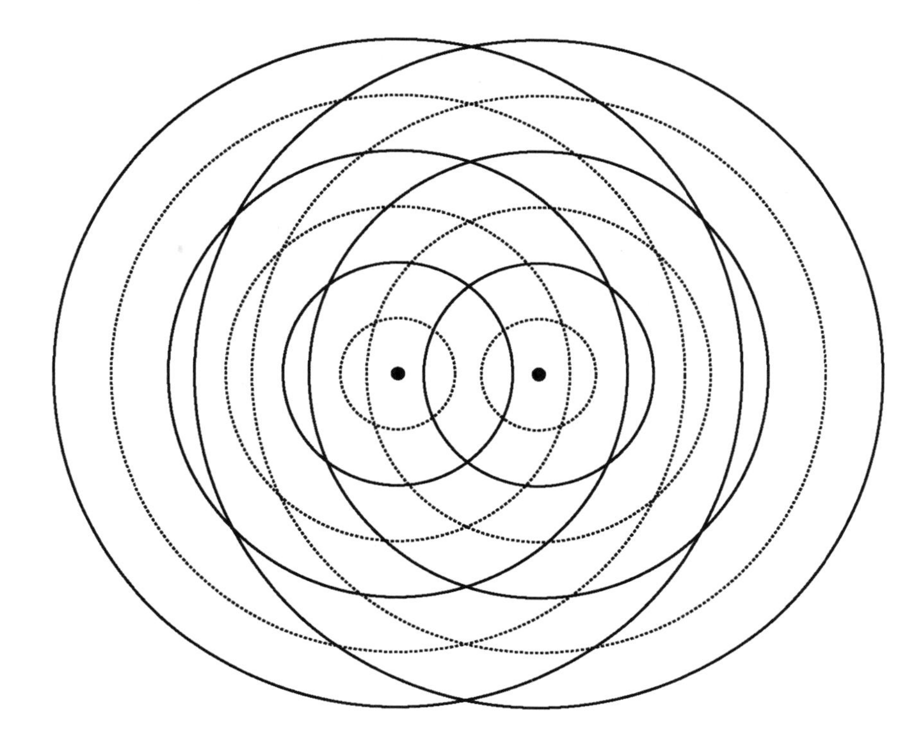

Name ______________________________ Date ____________ Class ____________

b. The two sources from the previous problem have been moved closer together. Use two different colors to draw the lines of constructive and destructive interference.

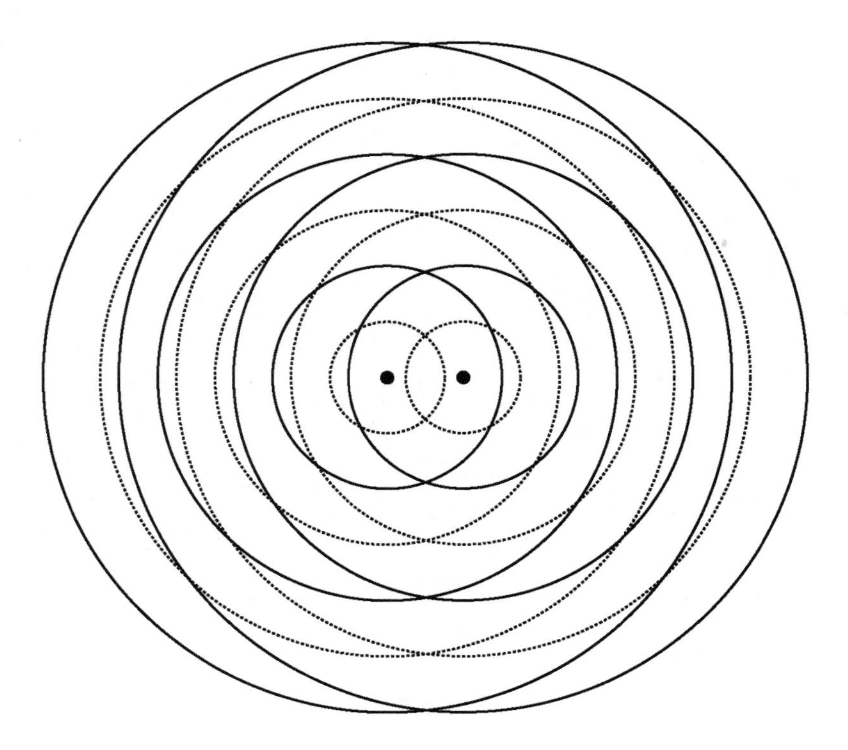

c. Describe how the lines of constructive and destructive interference change when the sources are moved closer together.

Name ______________________________ Date ____________ Class ____________

d. The wavelengths of the waves from the previous problems have been shortened. Use two different colors to draw the lines of constructive and destructive interference.

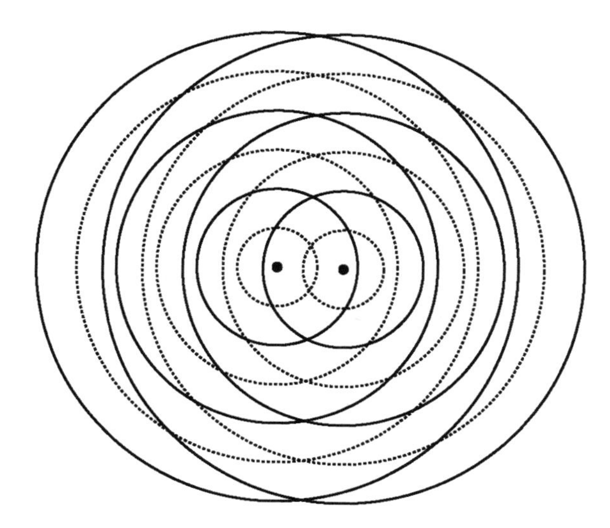

e. Describe how the lines of constructive and destructive interference change when the wavelength is shortened.

Name ______________________________ Date ____________ Class ____________

5. Use a ruler and compass to draw the wave crests and troughs if the wavelength of each source is 1 cm. Indicate the lines of constructive interference using one color pen or pencil and the lines of destructive interference using another color.

a. Roughly sketch what would be observed on the front of the viewing screen.

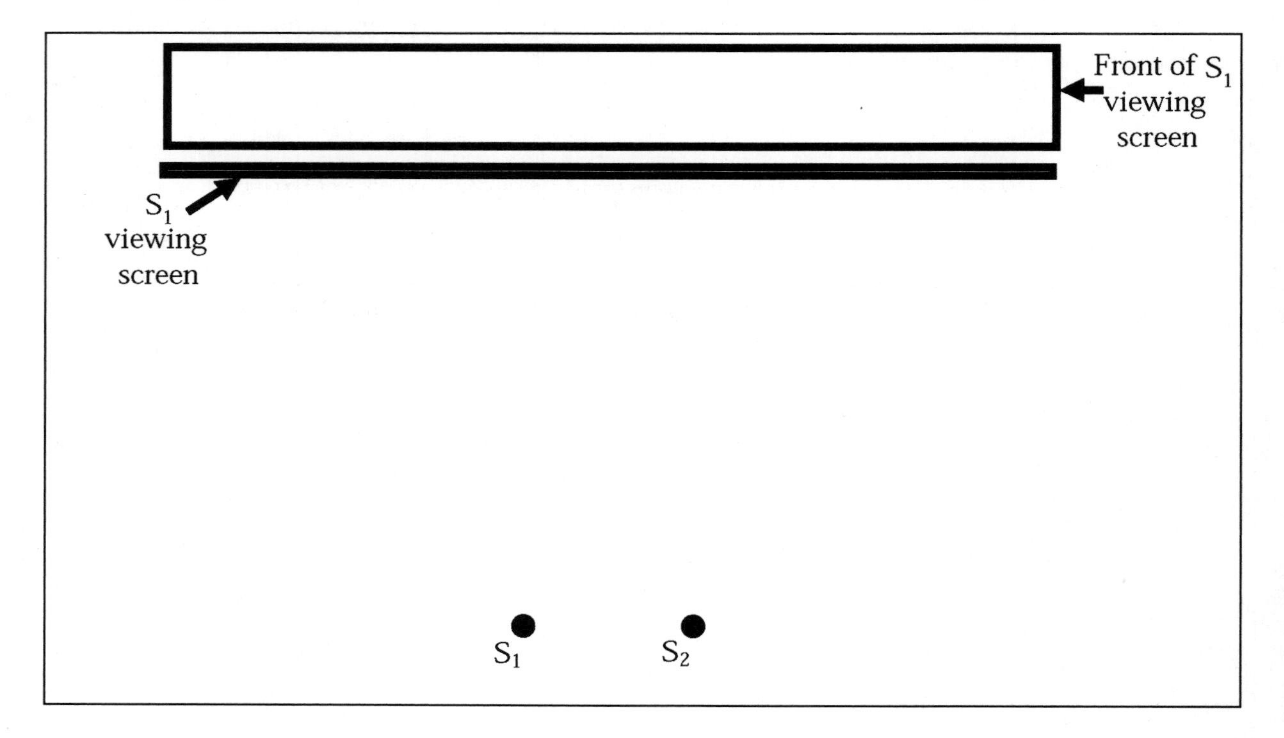

Name ______________________________ Date ____________ Class ____________

6. If the two sources are moved farther apart, how would you expect the interference pattern to change? Explain how you came to your answer.

   a. Use a ruler and compass to draw the wave crests and troughs if the wavelength of each source is 1 cm. The two sources are farther apart than they were in the previous question. Indicate the lines of constructive interference and destructive interference using different colors.

   b. Roughly sketch what would be observed on the front of the viewing screen.

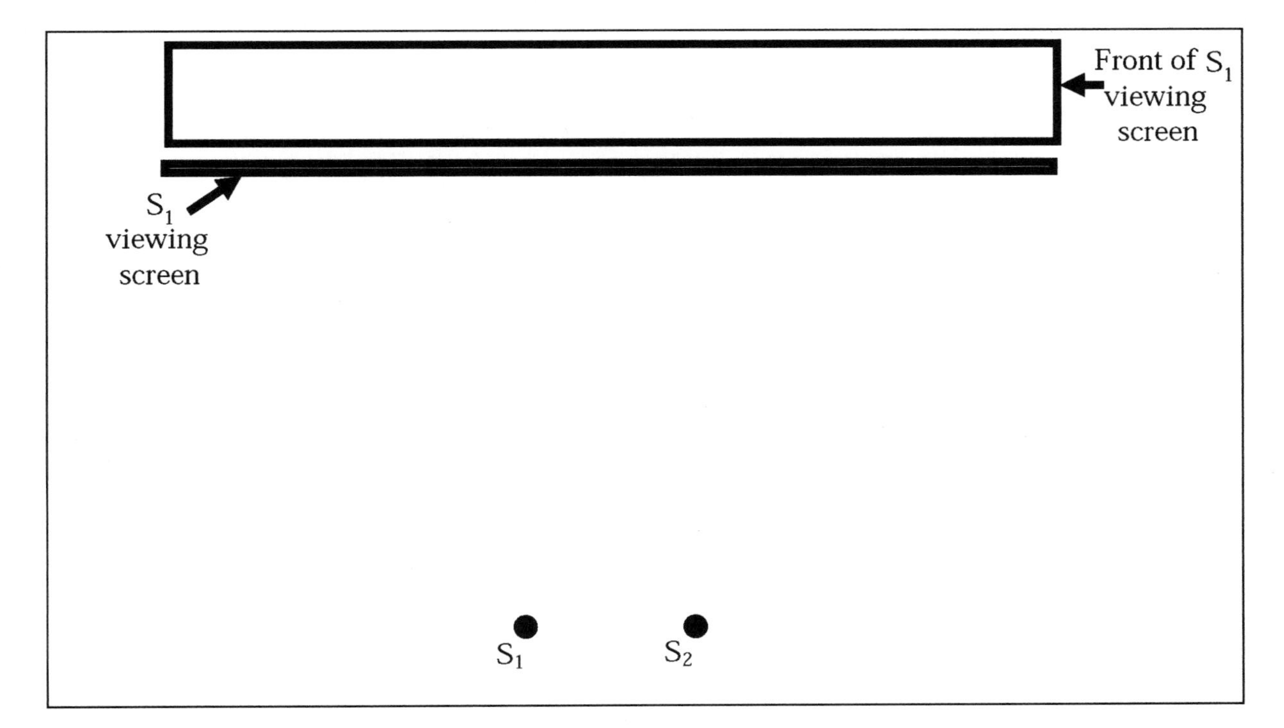

Name ________________________________ Date ____________ Class ____________

7. If the wavelength of the waves is reduced, how would you expect the interference pattern to change? Explain how you came to your answer.

   a. Draw the wave crests and troughs if the wavelength of each source is 0.5 cm. Indicate the lines of constructive interference and destructive interference using two different colors.

   b. Roughly sketch what would be observed on the front of the viewing screen.

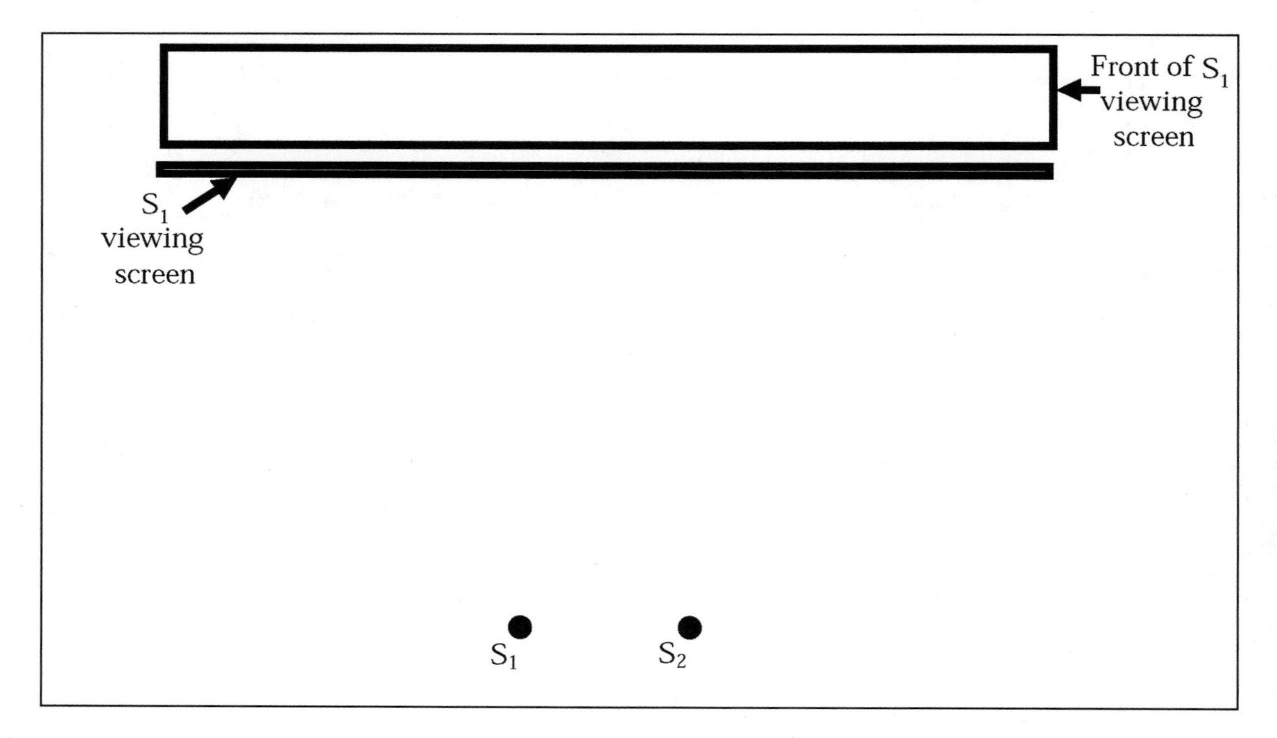

Name ______________________ Date ____________ Class ____________

8. When the distance between two sources increases, what happens to the distance between the maxima and minima observed on a screen?

9. When the wavelength of the waves decreases, what happens to the distance between the maxima and minima observed on a screen?

Name ______________________ Date __________ Class __________

Chapter 16

Geometry of Wave Path and Image Formation: Geometric Optics

1. A rectangular block is placed between a small, bright light bulb and a screen in an otherwise darkened room. Sketch the shadow that would appear on the front of the screen.

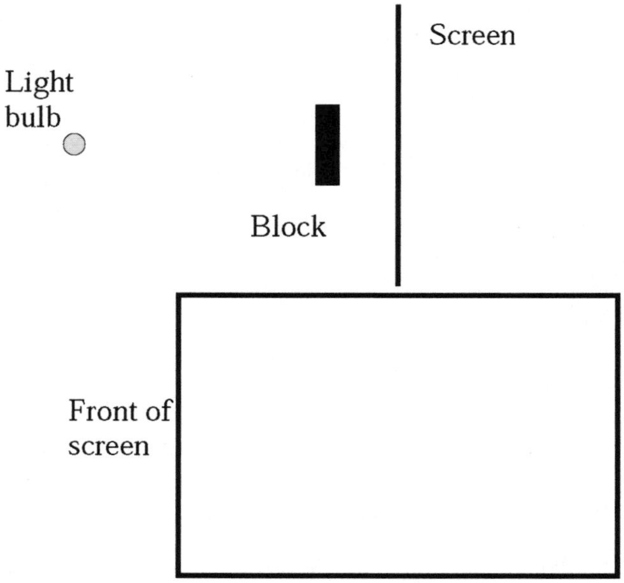

Screen

Light bulb

Block

Front of screen

Name ______________________ **Date** __________ **Class** __________

2. Sketch what would appear on the screen if the block were moved closer to the bulb.

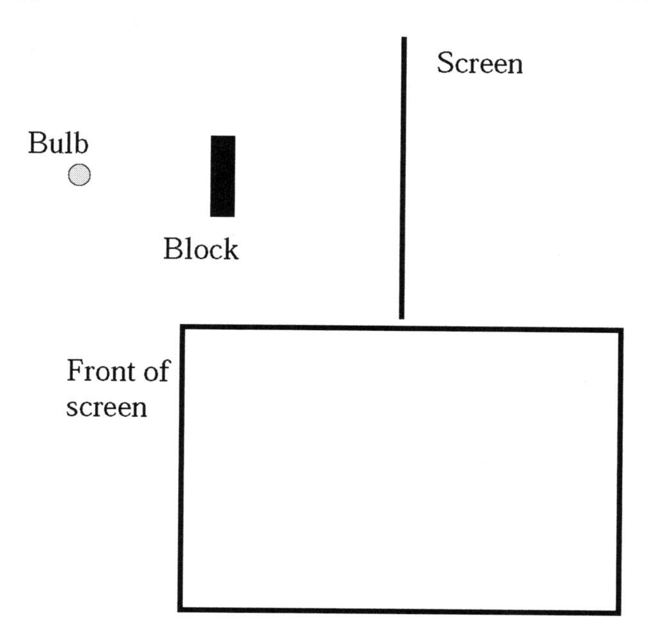

3. Sketch what appears on the screen when the bulb is moved farther back.

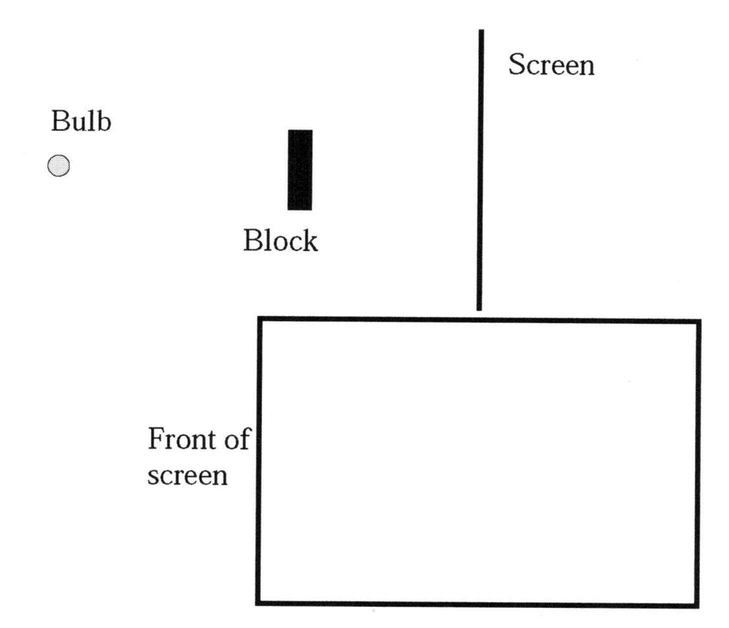

Name ______________________________ Date ______________ Class ______________

4. Explain to your friend what happens to the size of the shadow on the screen as the block or the bulb is moved back and forth.

5. An opaque screen with a tiny round hole in it is placed between a small, bright bulb and a viewing screen. Sketch what appears on the viewing screen.

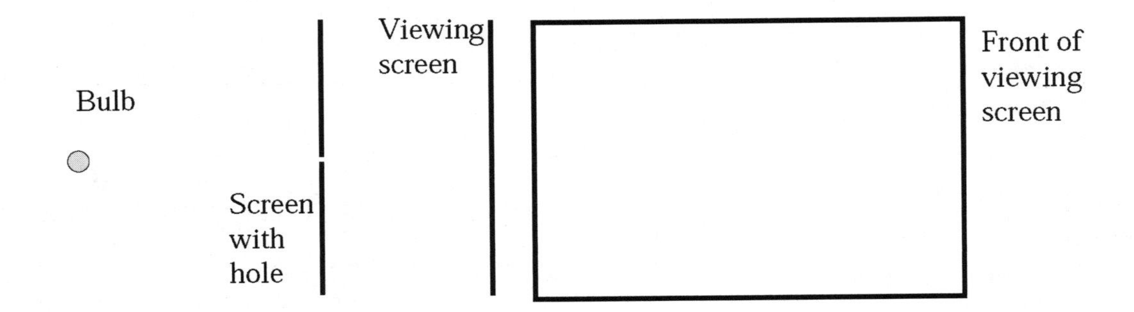

6. Two bright bulbs are placed to the left of an opaque screen with a tiny round hole and a viewing screen. Sketch what appears on the viewing screen.

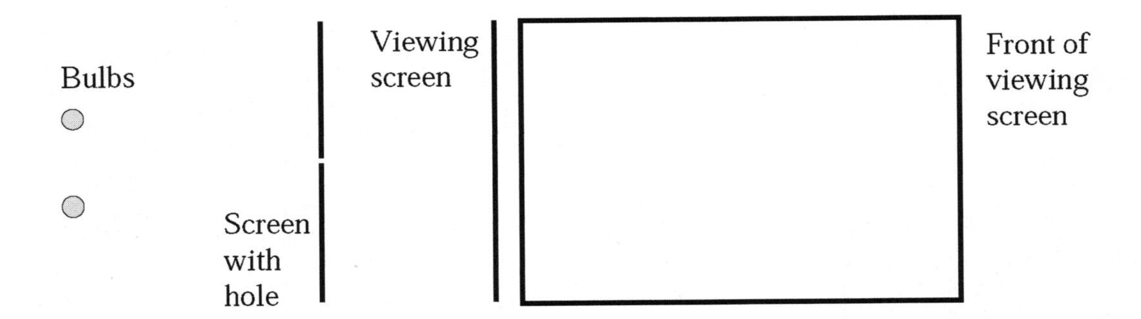

7. How would your sketch change if the two lights are moved closer together?

Name ______________________________ Date ____________ Class ____________

8. A long, thin filament is placed to the left of an opaque screen with a tiny round hole and a viewing screen. Sketch what appears on the viewing screen. (Hint: You can model the filament as several tiny bulbs lined up.)

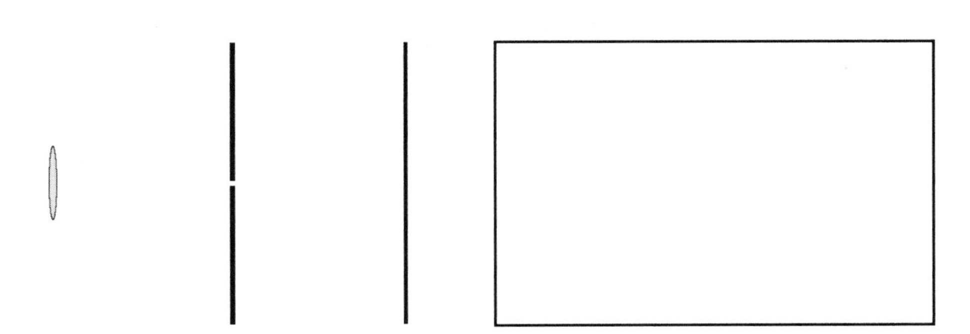

9. How would your sketch change if the filament moved toward the left relative to the hole (i.e., moved upward on the page)?

10. A screen with a rectangular horizontal hole cut in it is placed between a small, bright bulb and a viewing screen. Sketch what appears on the viewing screen.

11. Two bulbs are placed on one side of a screen with a rectangular horizontal hole cut in it. Sketch what appears on the viewing screen.

12. A long, thin filament is placed on one side of a screen with a rectangular horizontal hole cut in it. Sketch what appears on the viewing screen.

13. A long, thin filament is placed on one side of a screen with an "L" shaped hole cut in it. Sketch what appears on the viewing screen.

How do we locate an image?

14. Hold a pen or pencil in each hand about 8 inches from your face. Close one eye and try to bring the tips of the pencils together. Try this several times. How successful were you?

    a. Try it with both eyes open. Why is it easier with both eyes open?

    b. Hold one of the pencils about six inches from your face and alternate closing one eye and the other. What do you notice about the pencil as you switch eyes?

Name ______________________________ Date __________ Class __________

15. Your eyes are very good at telling which direction light rays are coming from. One way you can tell where an object is located is by comparing the different angles light from the object enters your eyes.

   a. Sketch the light rays that travel from the object to the center of each pupil.

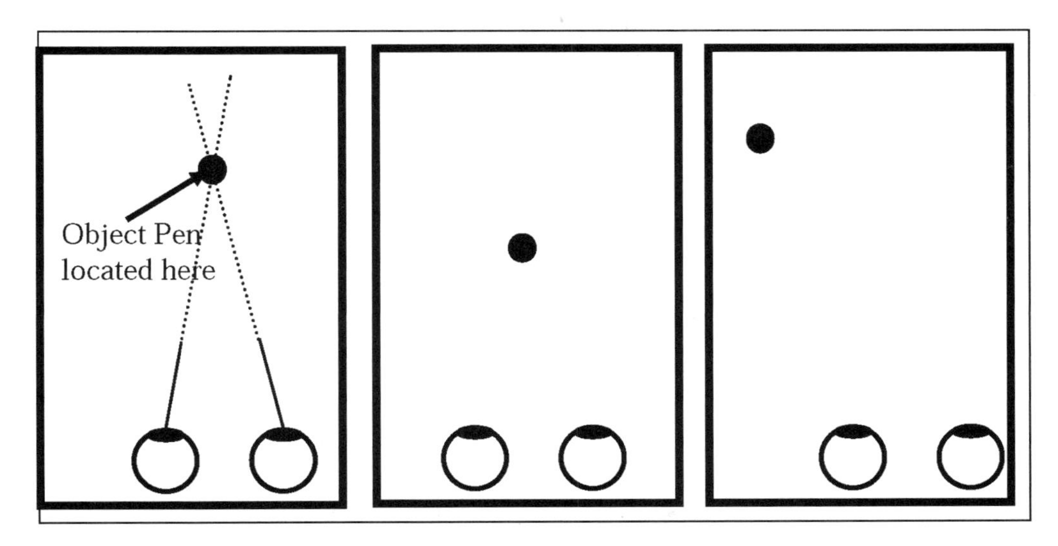

   b. Each eye perceives an incoming light ray from the directions shown. Sketch where the object is located. If you cannot locate the image, explain why not.

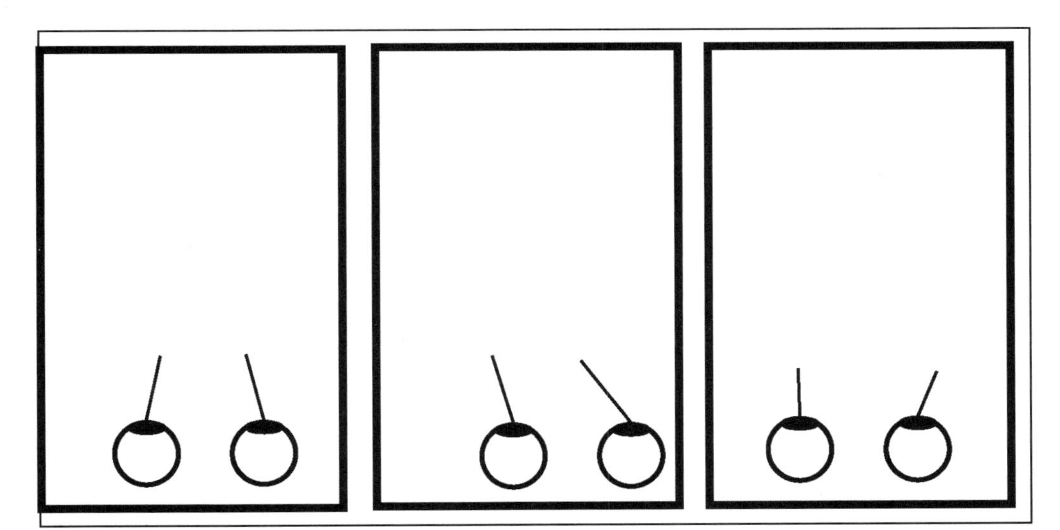

Name ______________________________ Date ____________ Class ____________

16. Use a ruler and protractor to draw the light ray from the object that reflects at the indicated points.

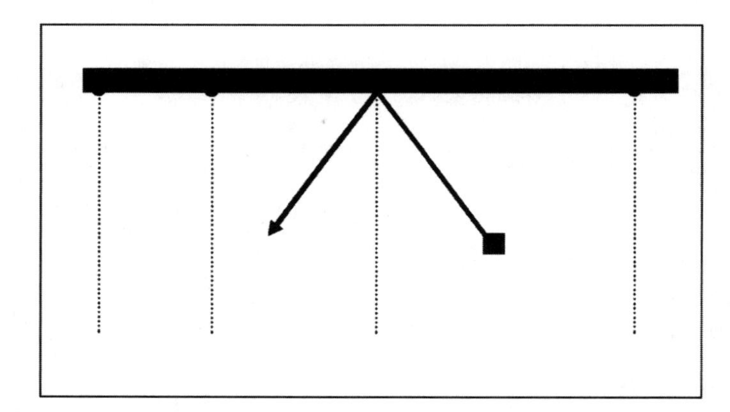

17. Draw the normal at each point on the mirror and sketch the light rays that reflect from the mirror at those points.

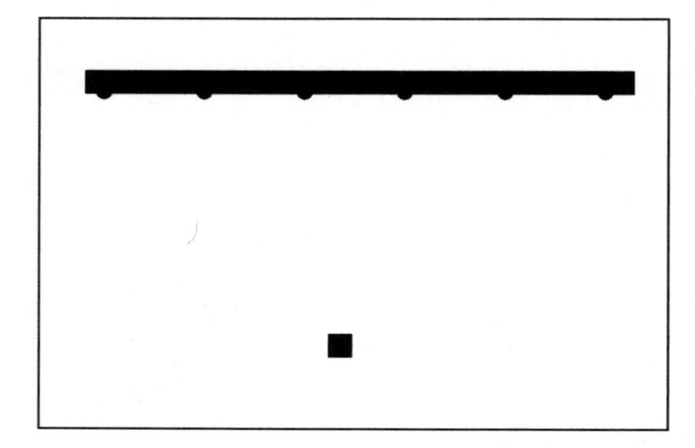

Name ______________________________ Date ____________ Class ____________

18. Draw the rays from the object that bounce off the mirror and enter the center of each pupil.

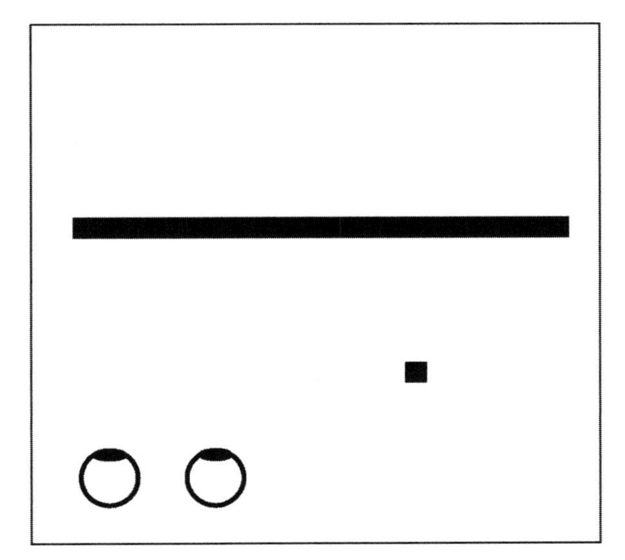

19. Based on the light rays entering the eyes, where does the image appear to the person?

20. Draw the rays from the object that bounce off the mirror and enter the center of each pupil. Based on the light rays entering the eyes, where does the image appear to the person?

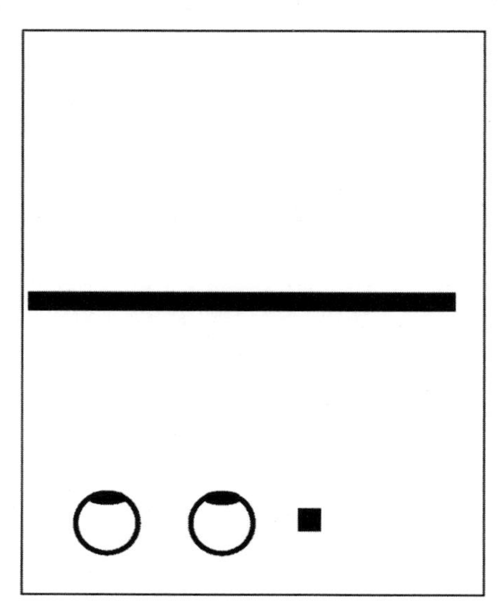

21. Draw the light rays to show where each observer sees the image in the mirror.

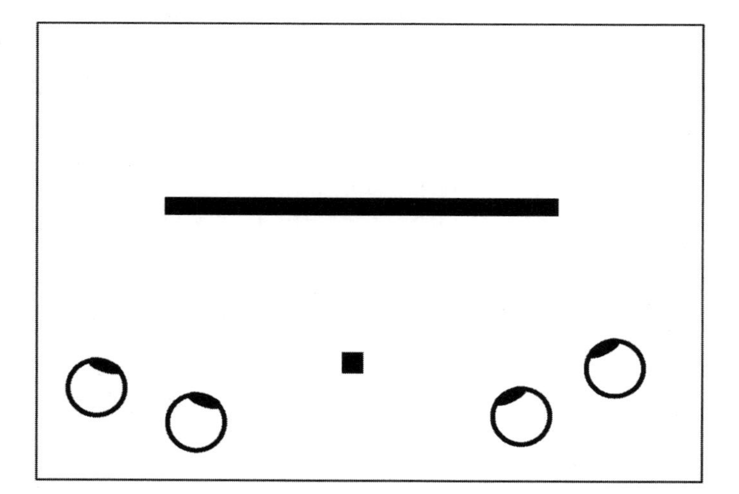

Name _______________________ Date ___________ Class ___________

22. For a mirror that is part of a sphere, it is very easy to find the normal at any point. A line drawn from the center point of the sphere (center of curvature) to the mirror is perpendicular to the surface at that point. Use a protractor and ruler to finish the ray diagram below.

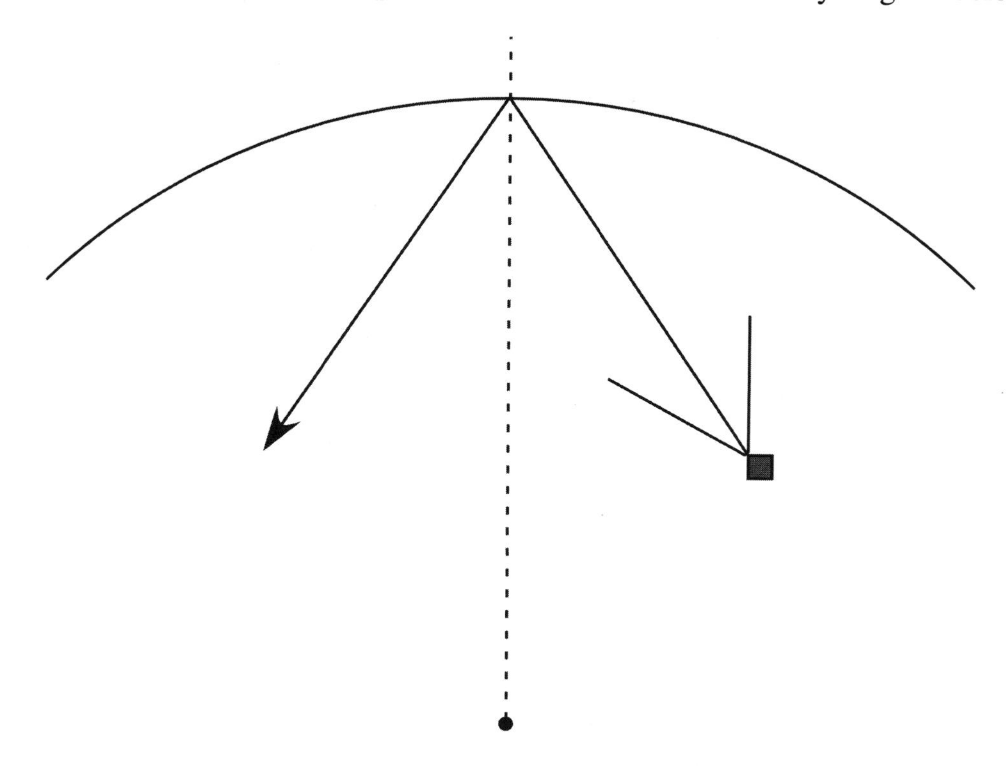

Name ______________________________ Date ____________ Class ____________

23. For a mirror that is part of a sphere, it is very easy to find the normal at any point. A line drawn from the center point of the sphere (center of curvature) to the mirror is perpendicular to the surface at that point. Use a protractor and ruler to finish the ray diagram below.

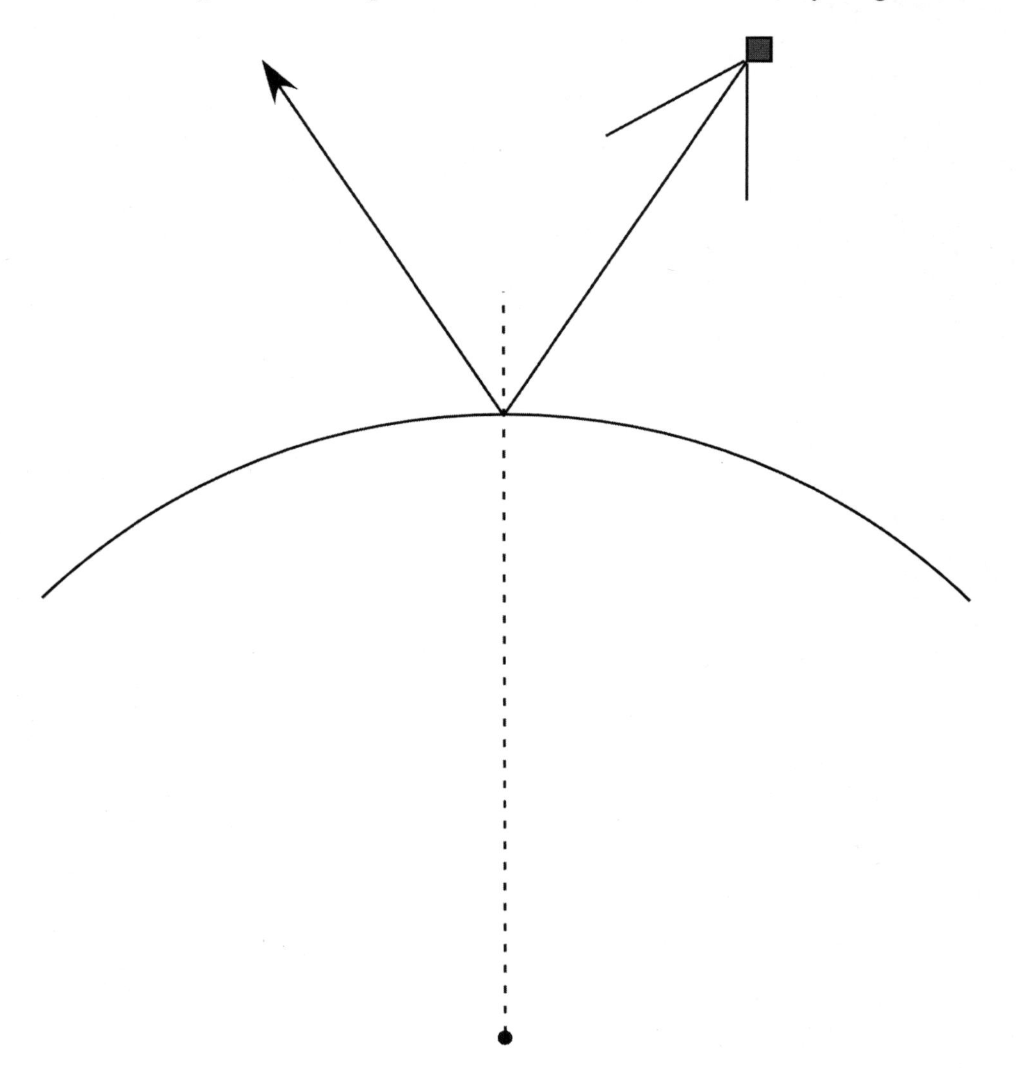

Name ______________________________ Date ____________ Class ____________

24. Locate where each observer sees the image by drawing the ray that leaves the object, bounces off the mirror and hits each eye at the center of the pupil.

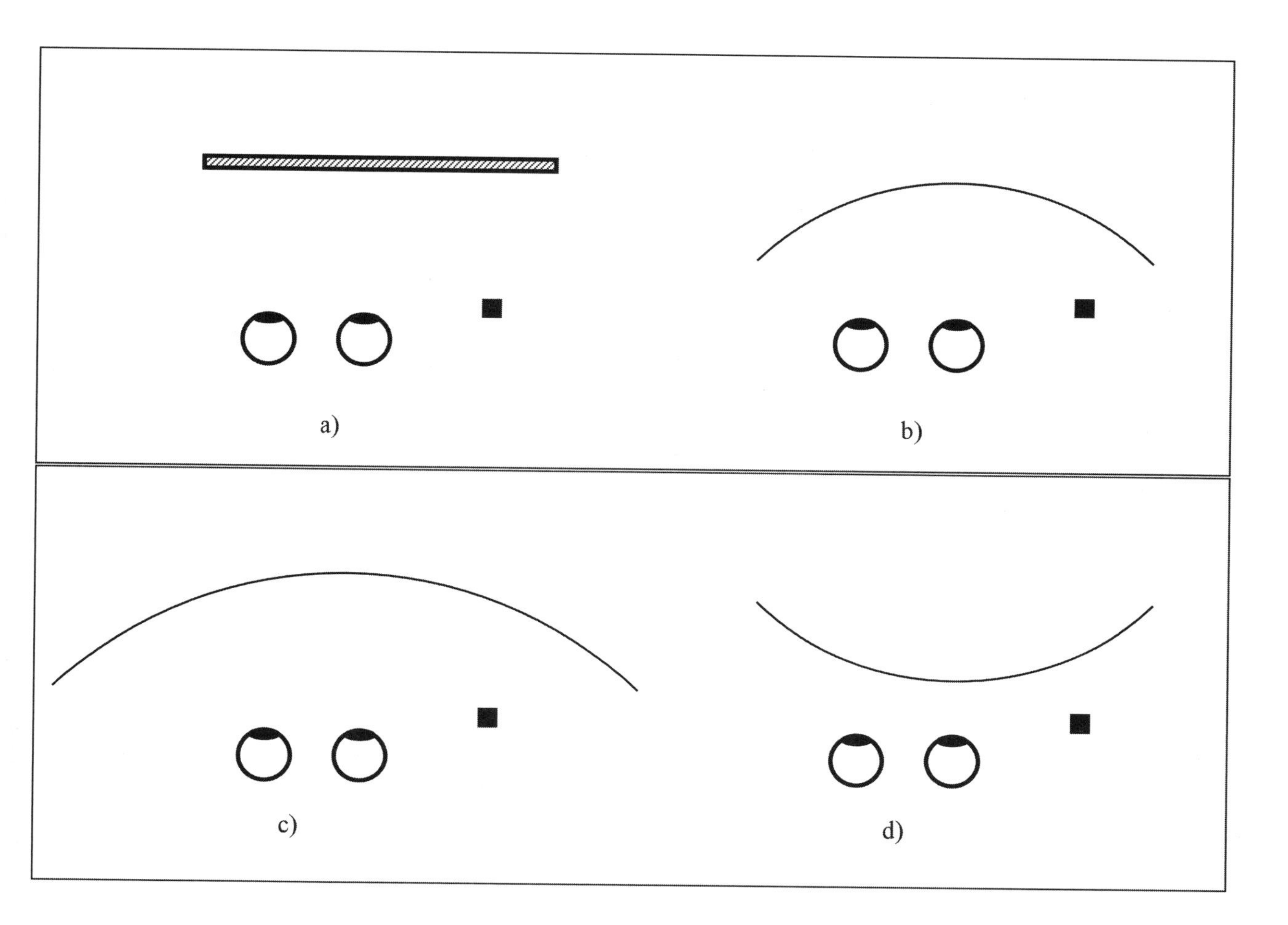

25. Rank the four sketches above (a-d) in terms of the distance between the image and object, starting with the shortest distance.

Name ______________________________ Date ____________ Class ____________

26. Sketch where you should place a concave mirror with a radius of 4 cm in order to see a real image of the object on the screen. Draw the rays to show the real image is on the screen.

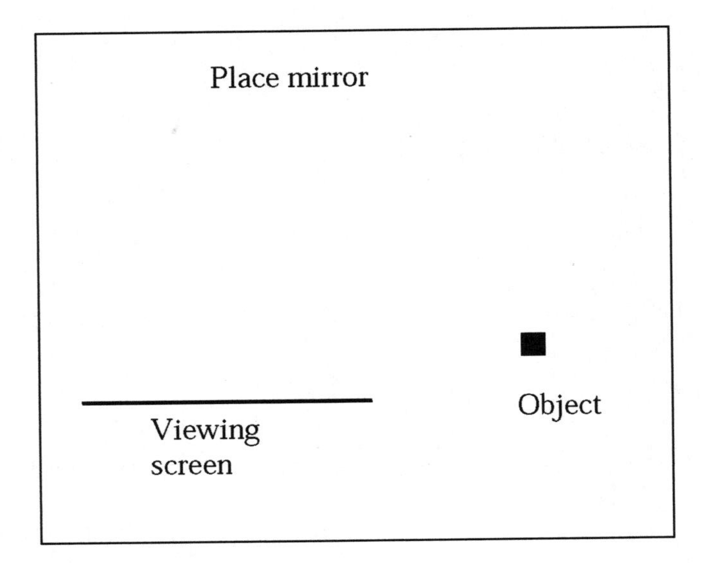

Refraction

27. Using a protractor and a ruler, sketch the reflected and refracted rays.

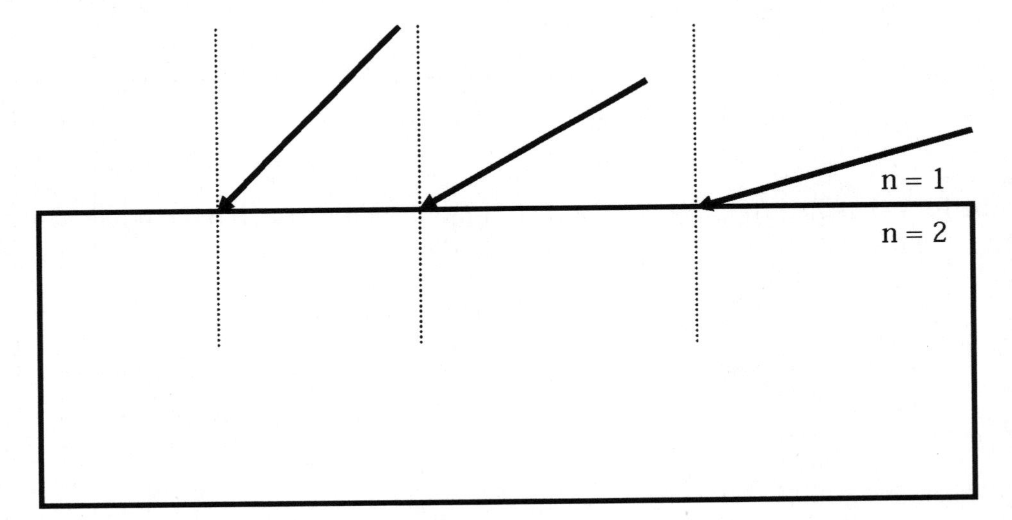

Name ______________________ Date ____________ Class ____________

28. Using a protractor and a ruler, sketch the reflected and refracted rays.

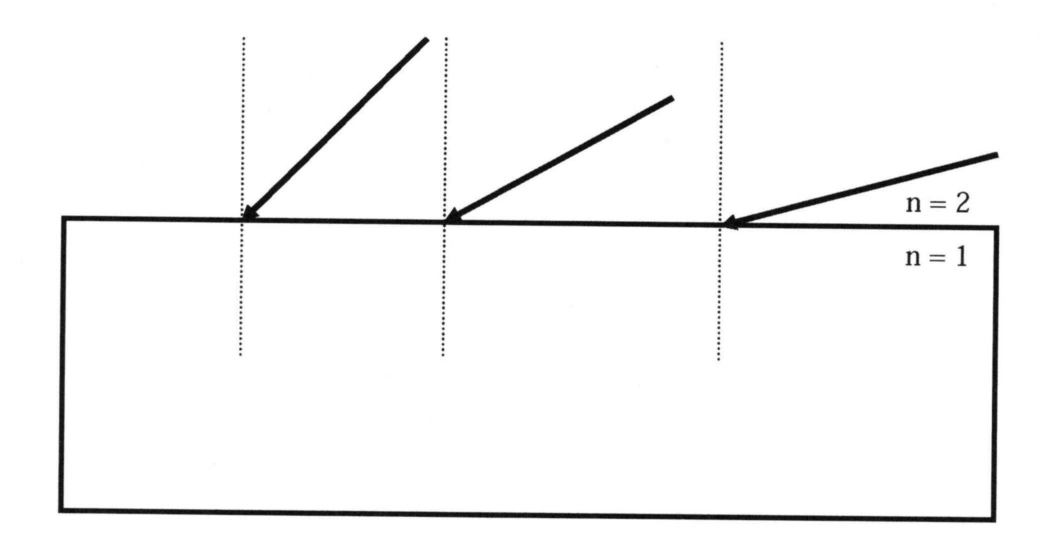

29. Where does each observer think the object in the water is located?

n = 1

n = 1.3

Name ______________________________ Date ____________ Class ____________

Chapter 17

Lenses: Optical Instruments

1. Use Snell's law to draw the light rays inside the denser medium.

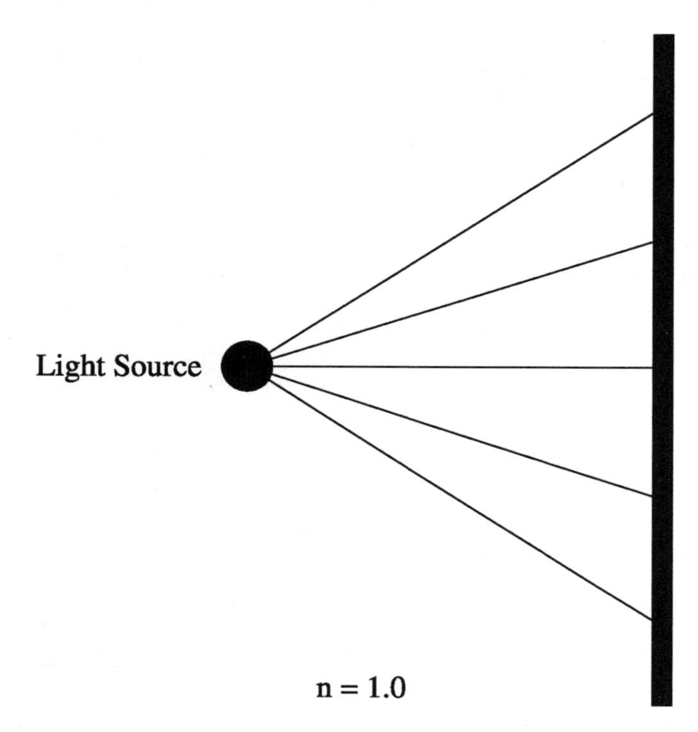

Name ______________________________ Date ____________ Class ____________

2. Use Snell's law to draw the light rays inside the denser medium. (Remember that the normal for a curved surface can be found by drawing a line from the center of curvature to the surface.)

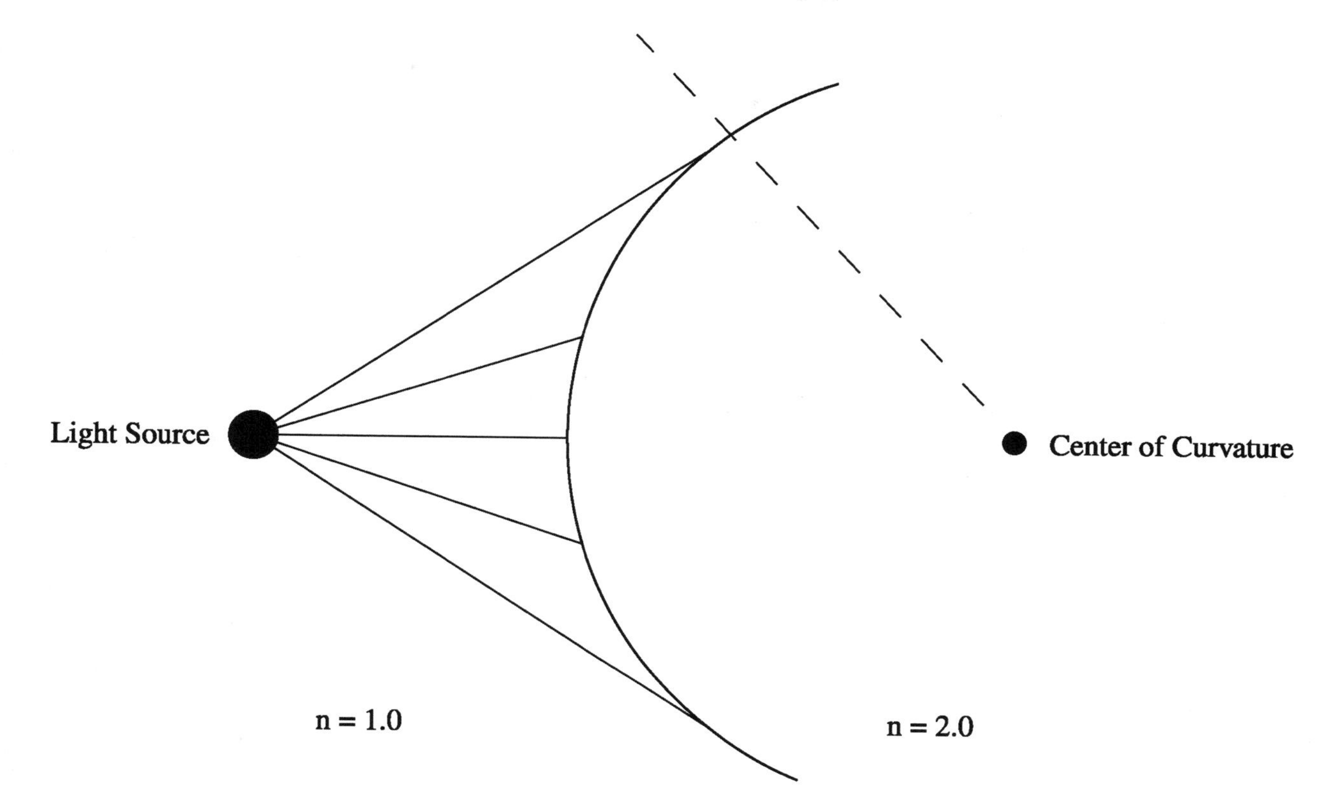

Name ____________________ Date __________ Class __________

3. Use Snell's law to draw the light rays inside the denser medium. (Remember that the normal for a curved surface can be found by drawing a line from the center of curvature to the surface.)

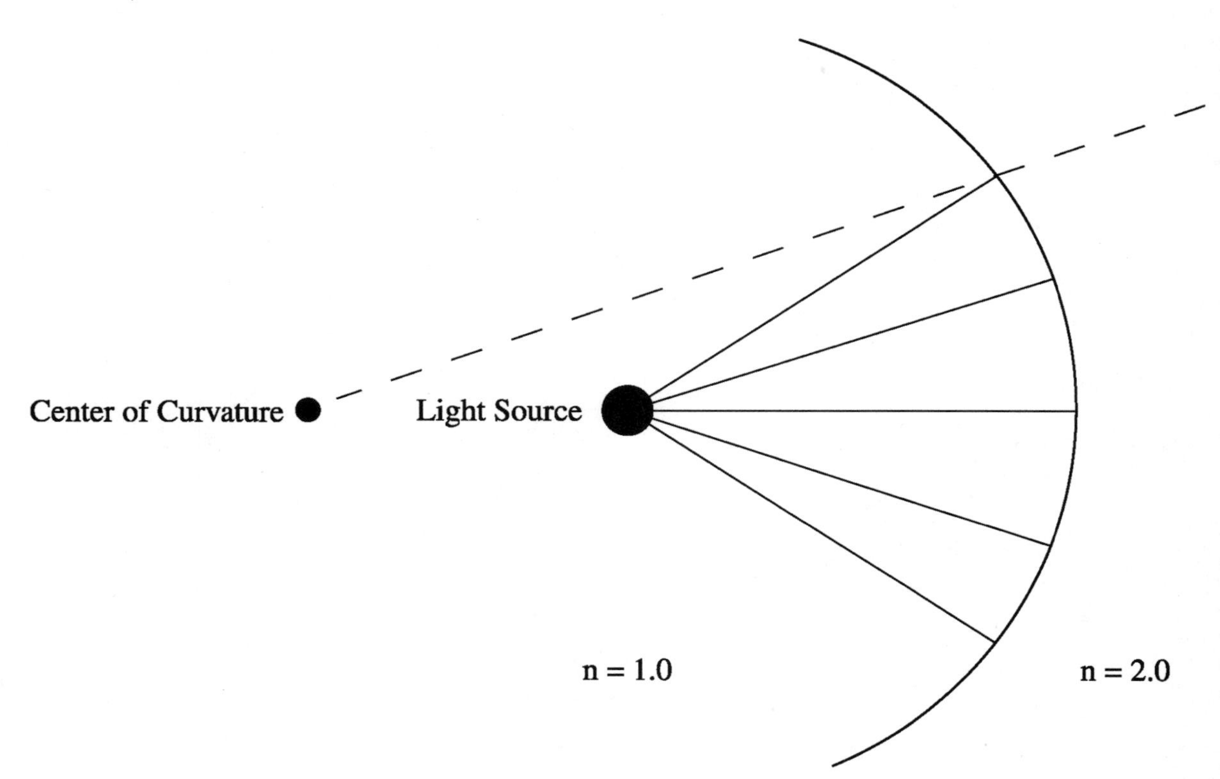

4. Circle the correct answer: The more outwardly curved the surface of a lens is, the **<more/less>** the light rays bend.

Name ______________________ Date ____________ Class ____________

5. Use Snell's law to draw the light rays inside the less dense medium.

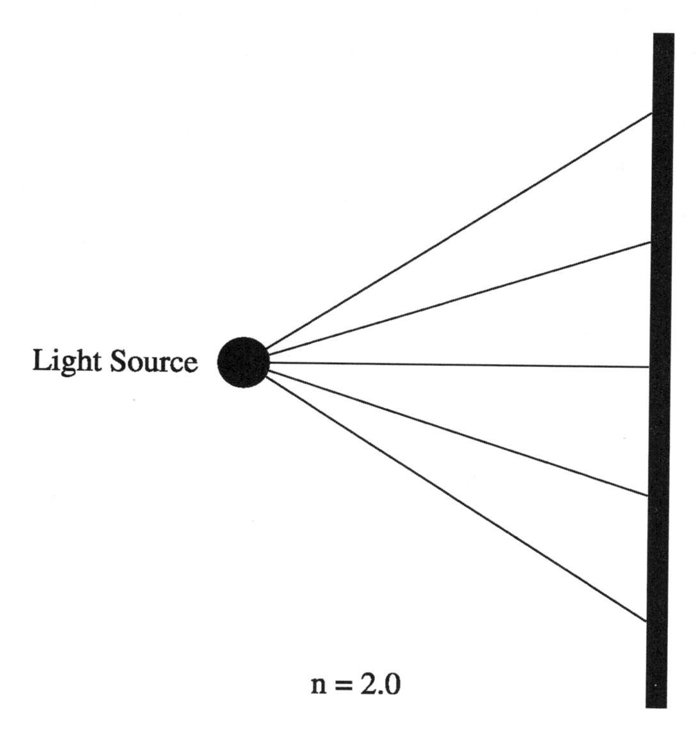

6. Use Snell's law to draw the light rays inside the less dense medium. (Remember that the normal for a curved surface can be found by drawing a line from the center of curvature to the surface.)

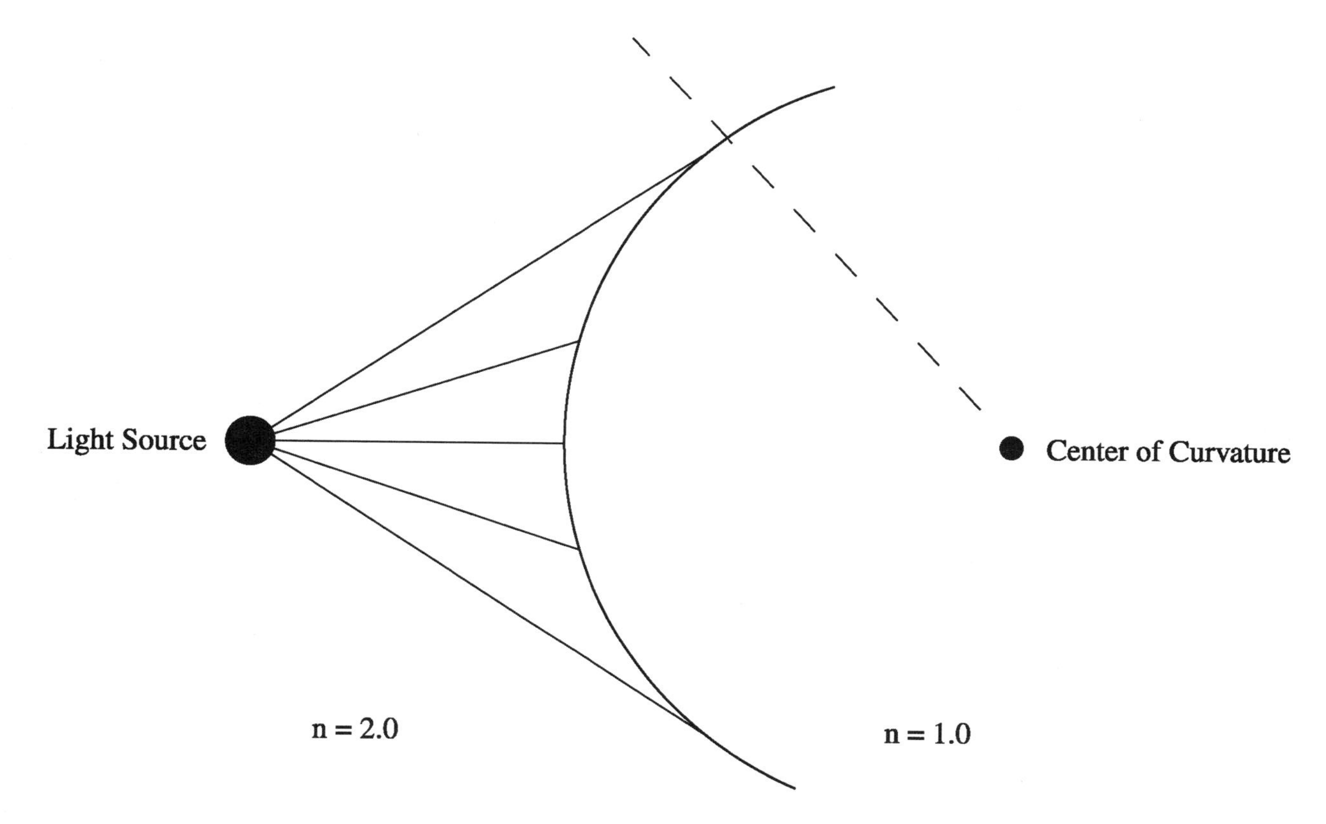

Name ______________________________ Date ____________ Class ____________

7. Use Snell's law to draw the light rays inside the less dense medium. (Remember that the normal for a curved surface can be found by drawing a line from the center of curvature to the surface.)

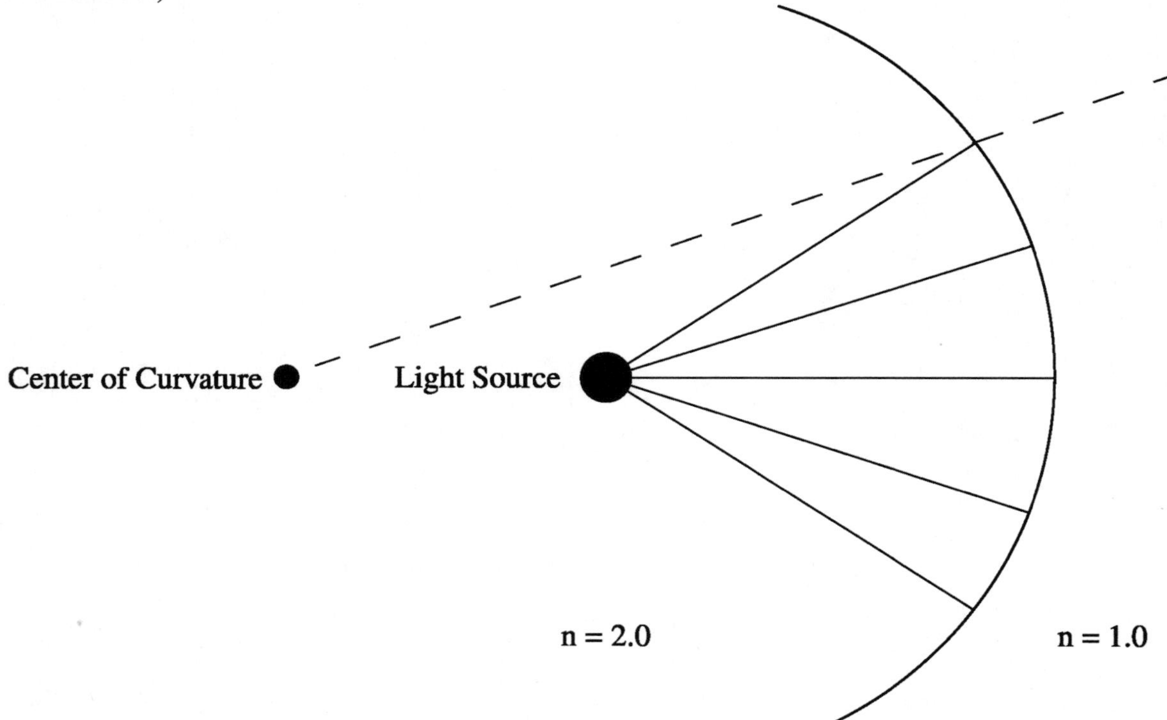

8. Circle the correct answer: The more inwardly curved the surface of a lens is, the **<more/less>** the light rays bend.

9. Fill in the table as to whether light rays bend inward (toward the center axis) or outward (away from the center axis). (COC stands for center of curvature.)

	Low Index to High Index	High Index to Low Index
Rays and COC same side		
Rays and COC opposite side		

Name ________________________ Date __________ Class __________

10. Locate the image of the small light source using the three rays already drawn. Once you've completed this, find the focal points of the lens and use the three primary rays to locate the image. Do these two results agree?

11. In order for your eye to see an object, the light from the object must form an image on your retina; a real image is formed on your retina. If a real image must be formed on the retina, how can we see virtual images?

12. Is the image formed by a pair of eyeglasses a real image or virtual image? Explain your reasons for choosing your answer.

Name ________________________ Date ____________ Class ____________

Chapter 18

Electrical Phenomena: Forces, Charges, and Currents

1. A positively-charged rod is brought near a small piece of aluminum foil hanging from a nylon thread. The aluminum foil moves toward the rod. When a negatively-charged rod is brought close, the aluminum foil moves toward the rod. Draw the charge configuration on the aluminum foil for each case.

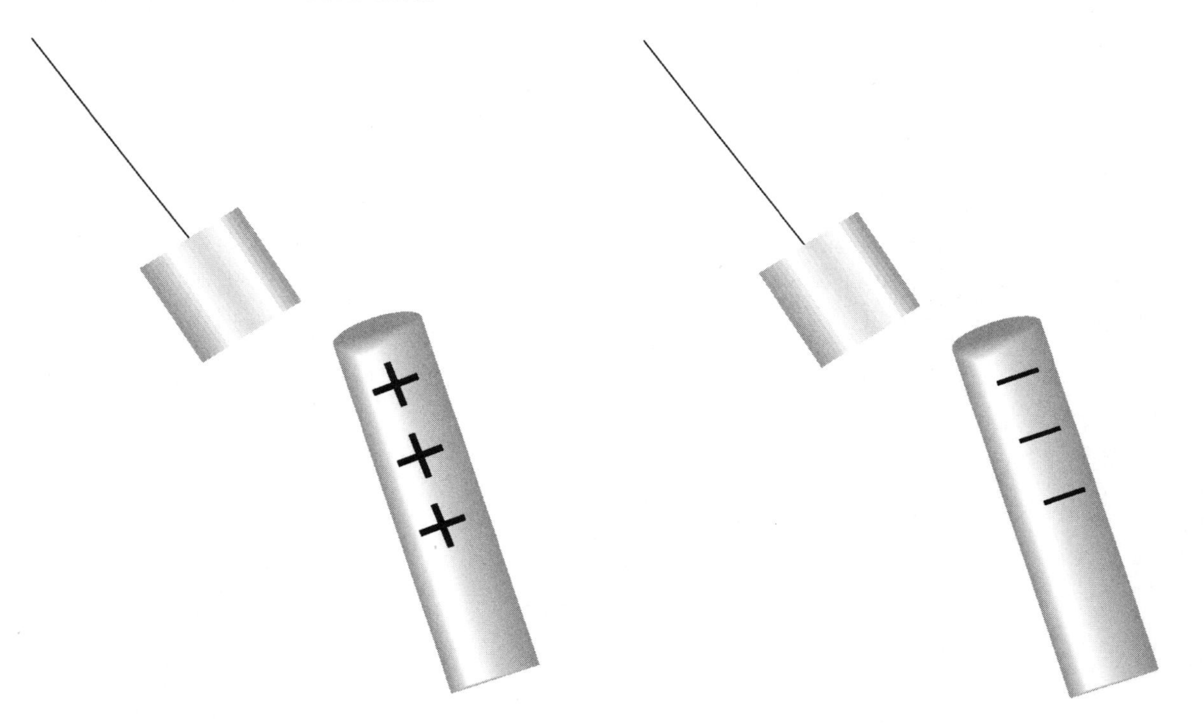

   a. What is the net charge on the aluminum foil? Explain your reasons.

   b. What is the net charge on the aluminum foil when the positive rod is brought near it?

   c. What is the net charge on the aluminum foil when the negative rod is brought near it?

Name ______________________________ Date ____________ Class ____________

d. If the foil was attracted to the positive rod but repelled by the negative rod, what is the net charge on the foil? Explain your reasons.

2. A negatively-charged piece of aluminum foil hangs from a nylon thread. Sketch the free body diagram for the foil as the rod is brought closer and closer.

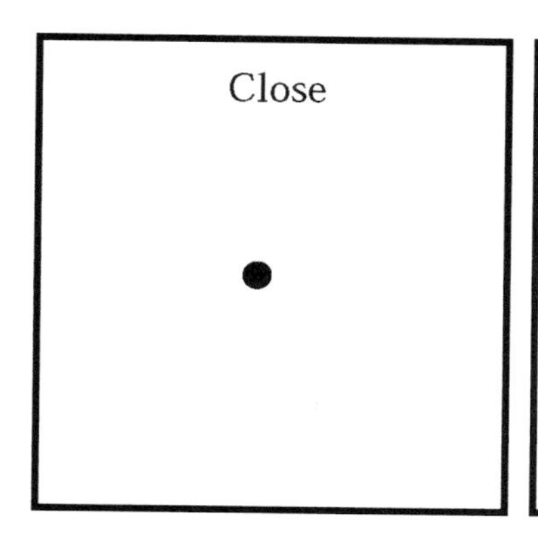

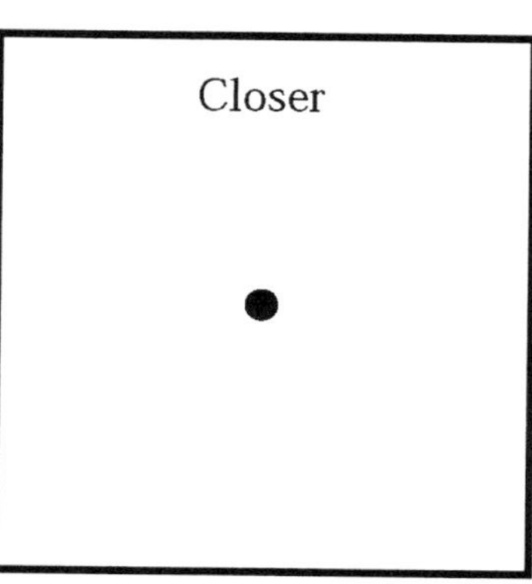

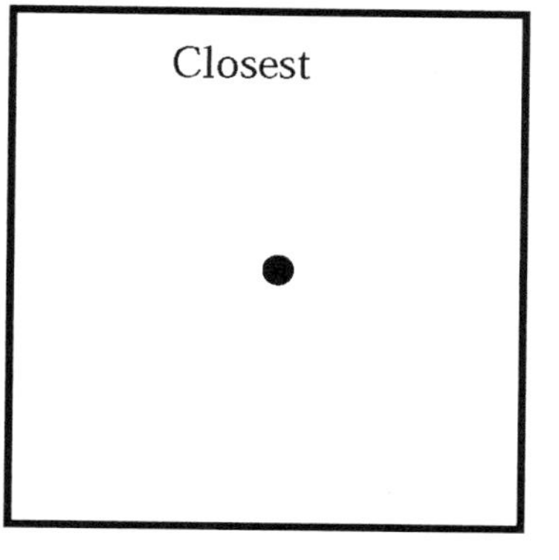

3. Two pieces of foil with identical charge are brought close to one another. Draw the free body diagram for each piece of foil.

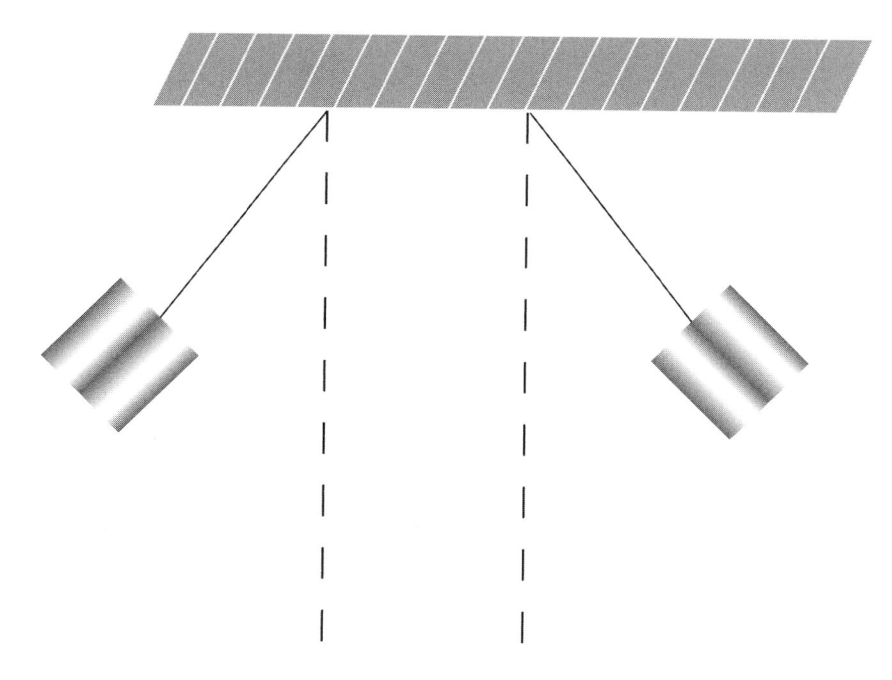

Name ______________________________ Date ____________ Class ____________

a. The charge on right-hand piece of foil is increased to three times the charge on the left-hand piece. Sketch how the two pieces will hang. Compare the angles at which they hang to the previous problem. How has the distance between the two pieces changed? Draw the free body diagram for both pieces.

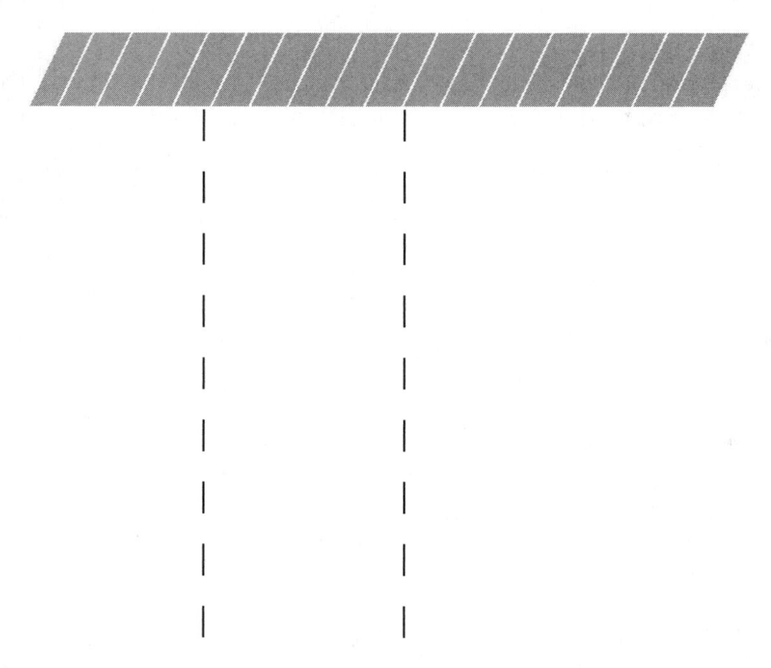

b. Explain why the electrostatic force each piece of foil experiences is the same for both pieces.

Name ______________________________ Date ____________ Class ____________

c. The right-hand piece is discharged. Sketch how the two pieces will hang. Compare the angles at which they hang to the previous problem. How has the distance between the two pieces changed? Draw the free body diagram for both pieces.

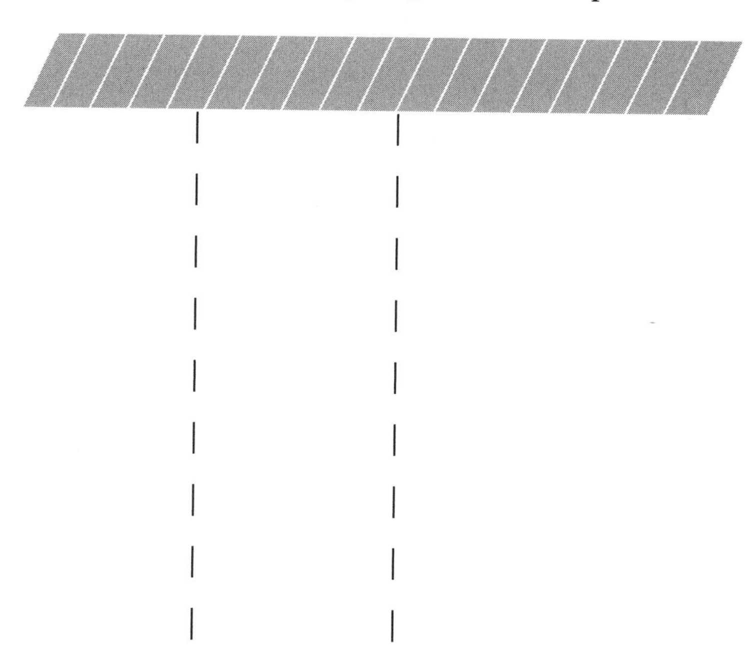

d. The weight of the left-hand piece of foil is increased to double the weight of the right-hand piece, but both pieces have the same charge. Sketch how the two pieces will hang. Compare the angles at which they hang to the previous problem. How has the distance between the two pieces changed? Draw the free body diagram for both pieces.

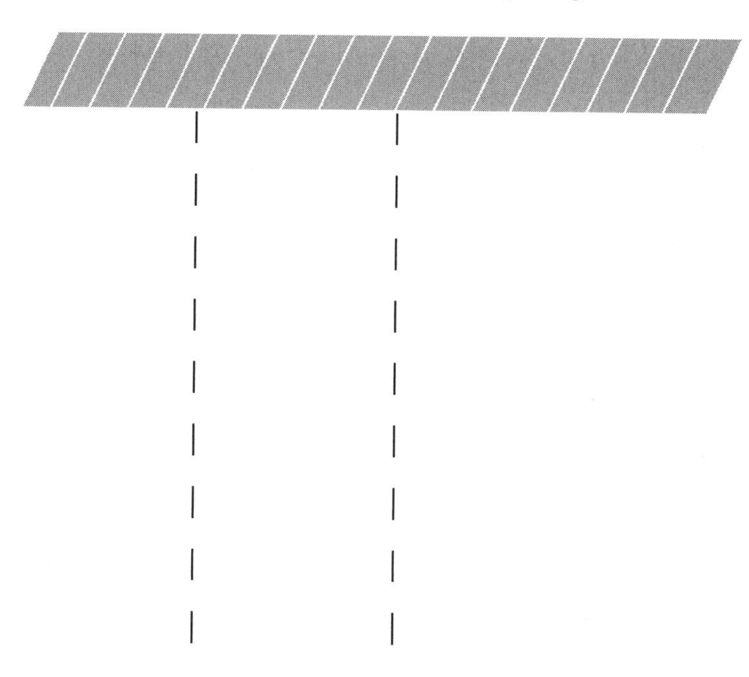

e. The weight of both pieces is increased to double their original weight and both still have the same charge. Sketch how the two pieces will hang. Compare the angles at which they hang to the previous problem. How has the distance between the two pieces changed? Draw the free body diagram for both pieces.

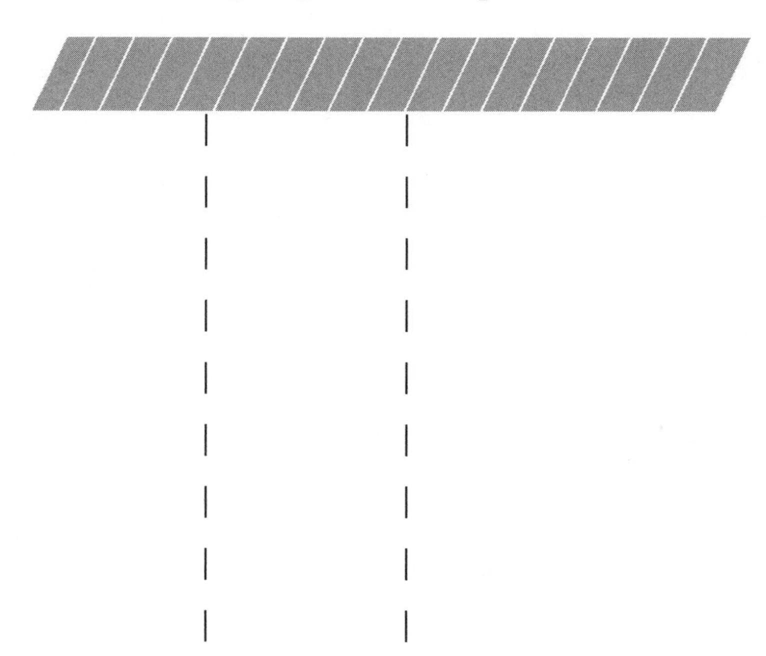

4. Calvin and Dennis are having a dispute about two charged pieces of foil. Calvin says, "If the pieces are attracted to each other, they both must be charged." Dennis says, "No, only one of the pieces of foil needs to be charged for them to attract."

   a. Give an example that proves Dennis is right. Include a sketch of the charge configuration on each foil.

   b. What can you say about the charges on the foil if the two pieces repel one another?

   c. Give an example that supports your argument. Include a sketch of the charge configuration on each foil.

Name ______________________________ Date ____________ Class ____________

d. Is it possible for the two pieces of foil to attract or repel one another due to electric forces if neither piece is charged? Explain your reasoning for your answer.

5. A positively- and negatively-charged pair of rods is brought close to (but not touching) opposite ends of a neutral metallic cylinder. Draw the charge configuration on the metal cylinder. (Assume the rods have the same amount of charge.)

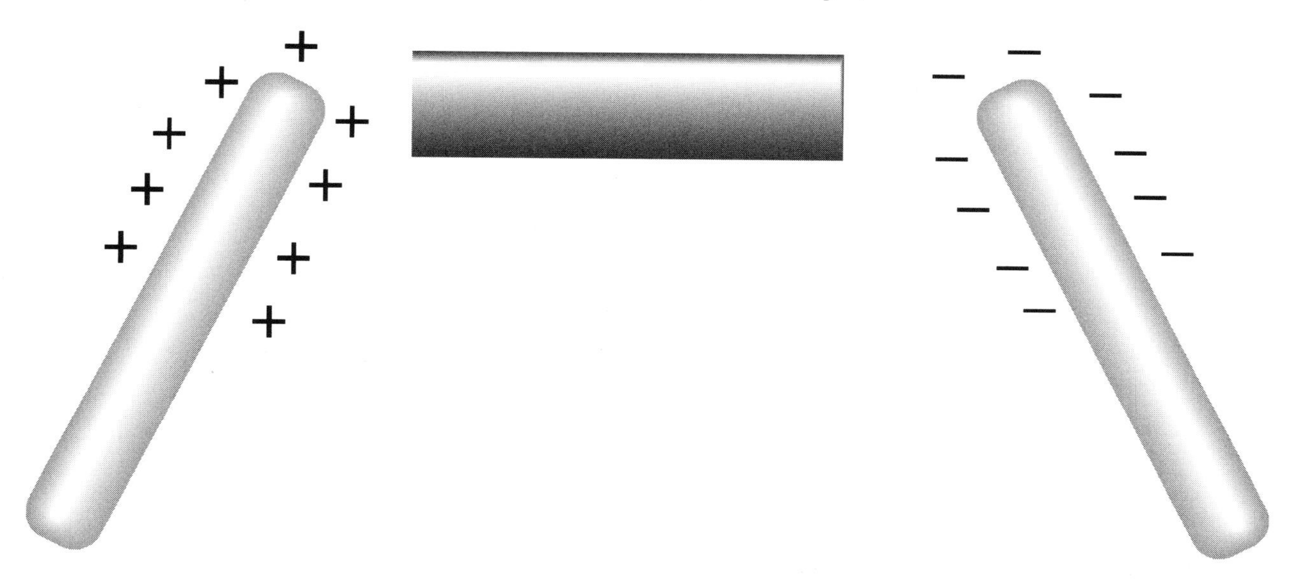

a. The charged rods touch the cylinder and then are moved far away. Draw the final charge configuration on the metal cylinder.

b. Two positively-charged rods are brought close to opposite ends of a neutral metallic cylinder. Draw the charge configuration on the cylinder.

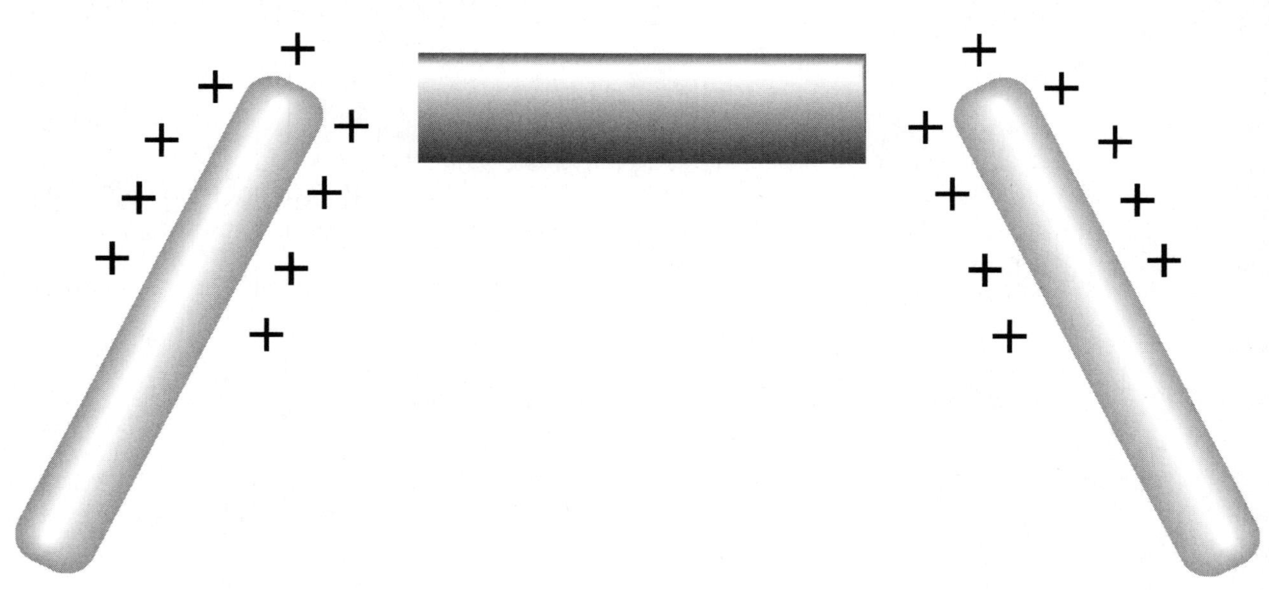

c. The positively-charged rods touch the cylinder and then are moved far away. Draw the final charge configuration on the metal cylinder.

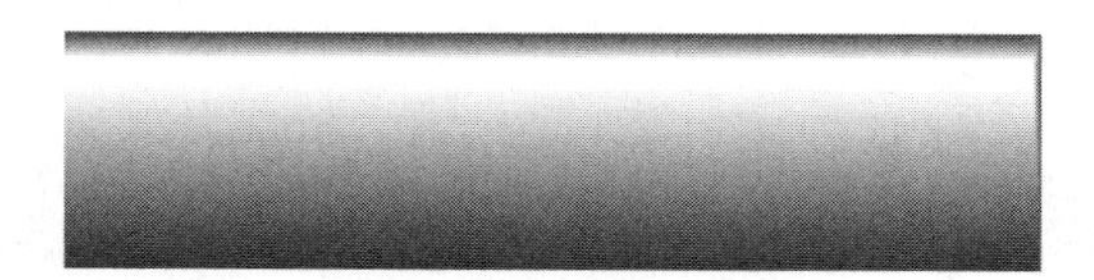

Name ____________________ Date __________ Class __________

Chapter 19

Electrical Field and Electrical Potential

1. The force exerted by a positive charge A on another positive charge B located 0.4 cm apart is $F_{A\ on\ B} = 7 \times 10^{-3}$ N.

   a. What would the force be when the charges are 0.5 cm apart?

   b. What would the force be when the charges are 4.0 cm apart?

   c. In general, if the force exerted by A on B is F_1 when they are d_1 apart, what is the force F_2 (exerted by A on B) when they are d_2 apart?

   d. What is F_2 (in terms of F_1) if d_2 is twice as big as d_1?

   e. What is F_2 (in terms of F_1) if d_2 is one third as big as d_1?

Name ______________________________ Date ____________ Class ____________

2. Draw the vector for the force exerted on a positive charge by the central positive charge at each dot on the diagram. The length of each vector should be proportional to the magnitude of the force. One force vector is drawn for scale.

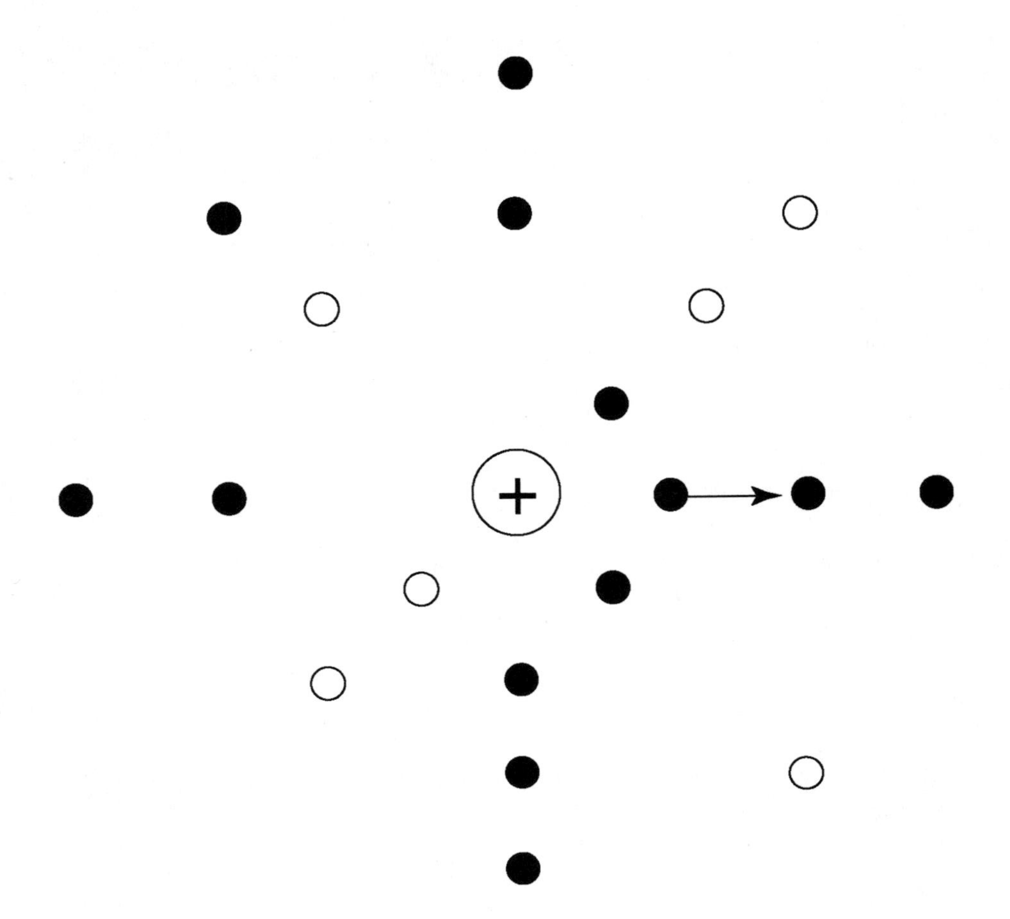

a. How would the electric field vectors look compared to the force vectors?

b. What benefits are there to using electric fields rather than just calculating the electrostatic forces?

Name ______________________________ Date ____________ Class ____________

3. Draw the electric field vectors for a negative charge at each point indicated.

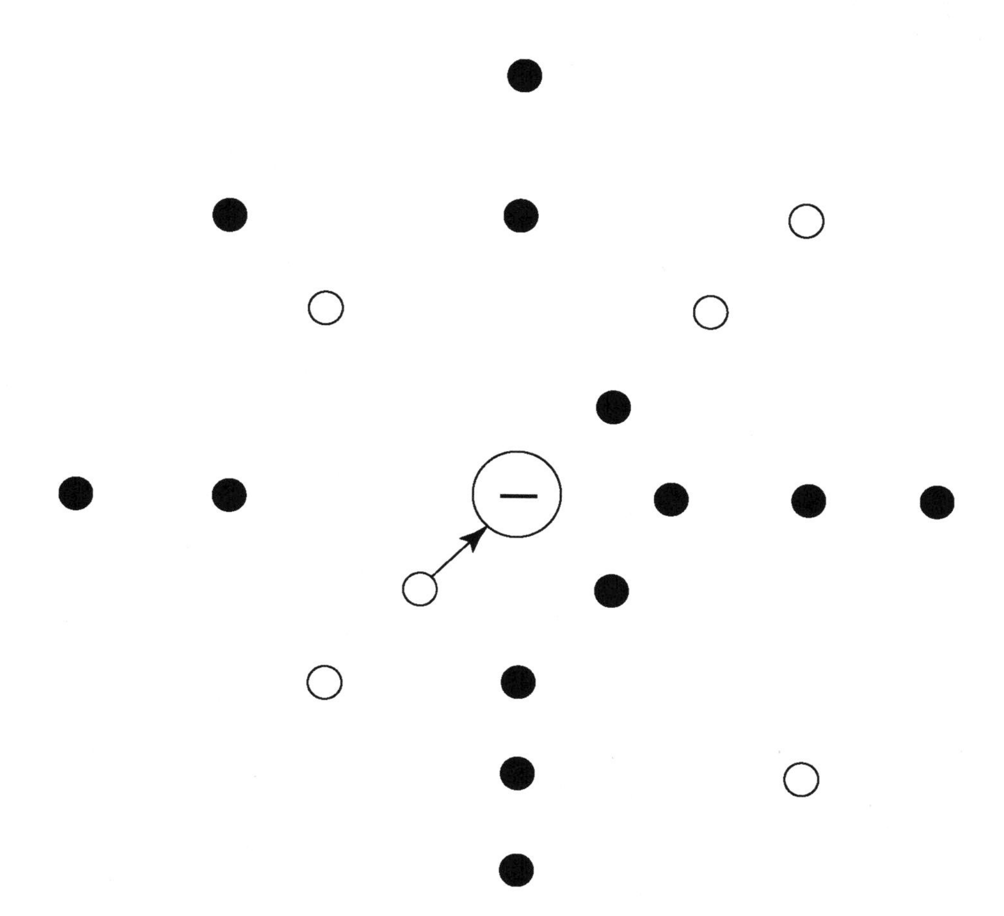

a. Use a different color to sketch the electric field lines.

b. How can you determine whether the charge is positive or negative from the field lines alone?

Name ______________________ Date ____________ Class ____________

4. Draw a vector representing the force experienced by each charge. The length of the force vectors should be proportional to the force.

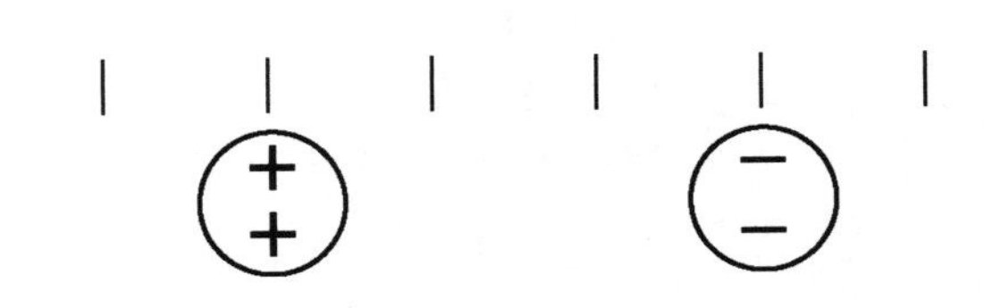

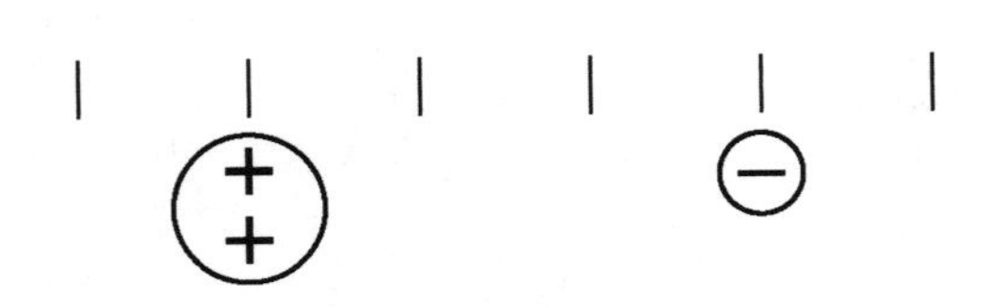

5. Draw the force vector experienced by the indicated charge due to each of the other charges, and label which charge is exerting that force. Use a different color to graphically add the vectors to find the net force experienced by the charge.

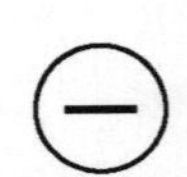

Name ______________________ Date __________ Class __________

6. Draw a vector representing the electric field at each dot due to each charge and use a different color to show the sum of the two fields.

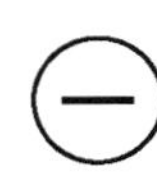    

7. How do the field vectors in the previous problem compare with the force vectors in the problem before that? Explain why the two are related.

8. A positive charge is "glued" at one location so it cannot move. You are pushing a second positive charge toward the first.

   a. What happens to the force experienced by your charge as they get closer together?

   b. Are you doing positive or negative work by pushing the charge? Explain your reasoning.

c. What type of energy is your work converted into?

d. What happens to the amount of work you do moving the charge 1 cm as the two charges get closer together? Explain your reasoning.

e. The electrostatic force is a conservative force. What does this mean? (Hint: Look back in Chapter 6.)

9. It is useful to look at the parallels between a gravitational field and an electric field. Near the Earth's surface, the gravitational field is roughly a constant field. The following series of sketches (seen on the next page) shows a baseball, initially at rest, that is then dropped. Fill in the energy bars for the baseball/Earth system next to each sketch.

   a. Explain why you know the magnitude of the force on the ball does not change as the ball falls in a constant gravitational field.

   b. Is any work done on the ball/Earth system?

Name ______________________ Date ____________ Class ____________

c. Does the total energy of the ball/Earth system change as the ball falls?

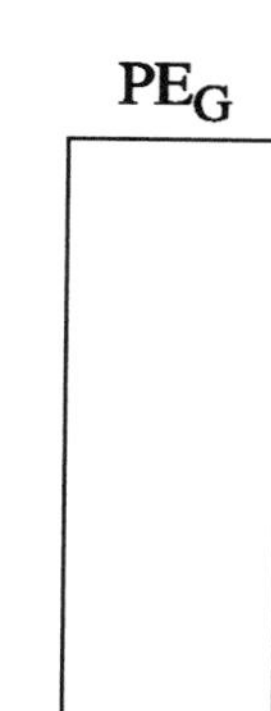
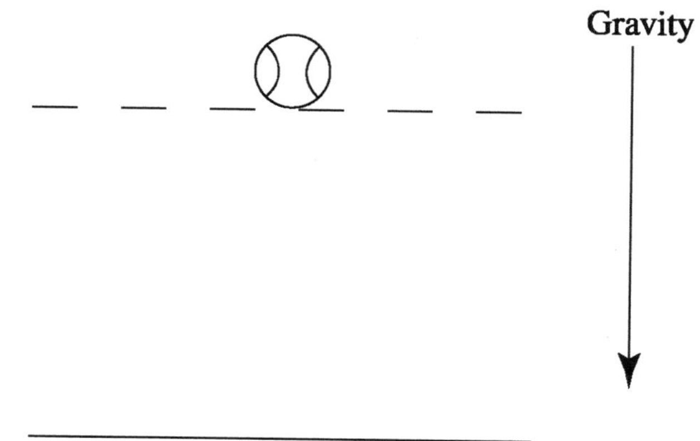

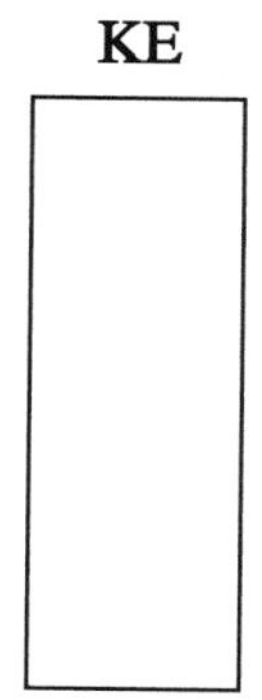

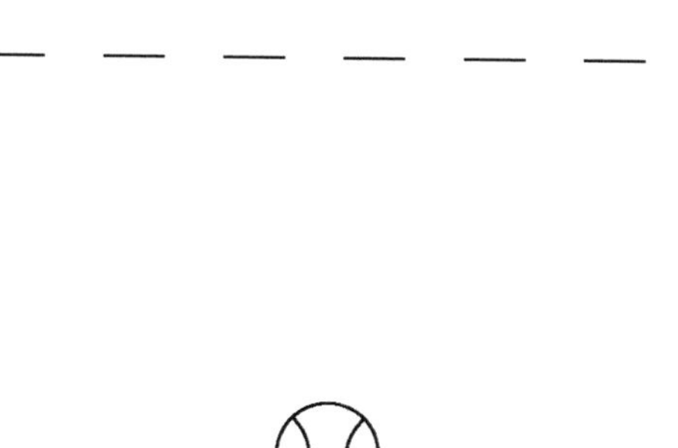

Name ________________________________ Date ____________ Class ____________

10. The sketches below show a positively-charged particle in a constant electric field. Fill in the energy bars for the series of sketches. Assume the charge is initially stationary. The electric field and the positive charge are part of the same system.

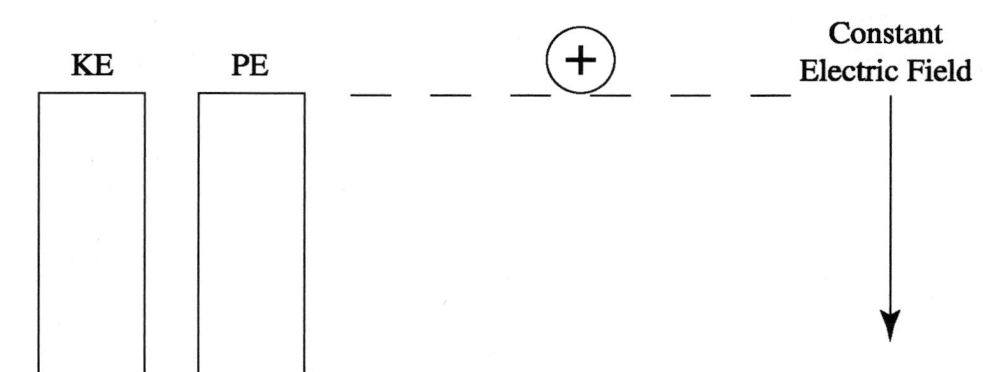

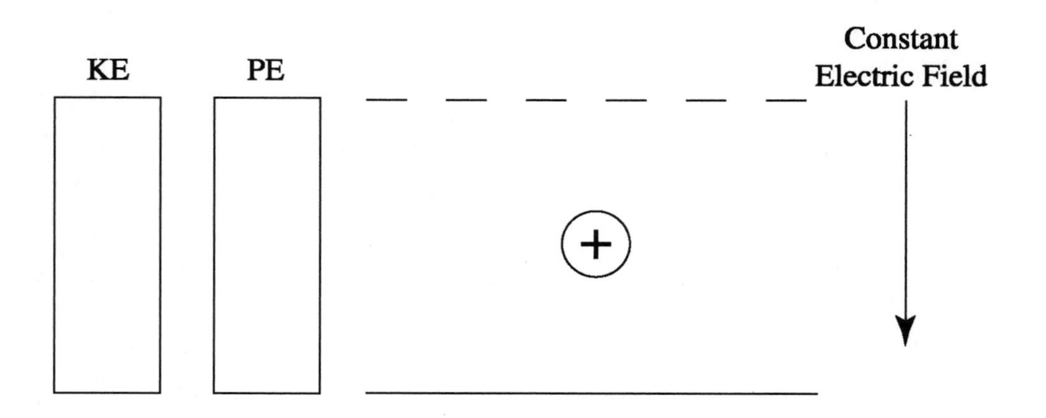

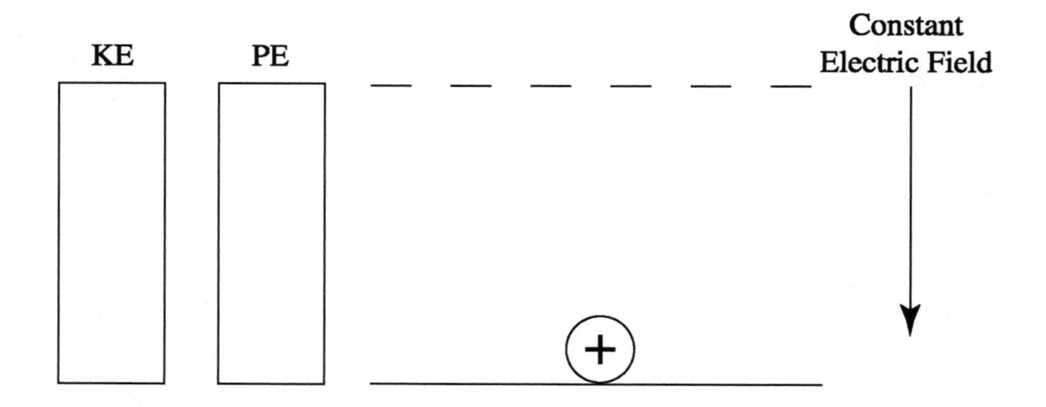

Name ______________________ Date ____________ Class ____________

11. Does the magnitude of the force on the positive charge change as it moves?

a. Is any work done on the positive charge/electric field system?

b. Does the total energy of the charge/electric field system change?

Name ______________________________ Date ____________ Class ____________

12. A ball is thrown straight up in the air with some initial velocity v_0. Fill in the energy bars as the ball rises. (The dotted line indicates the maximum height the ball reaches.)

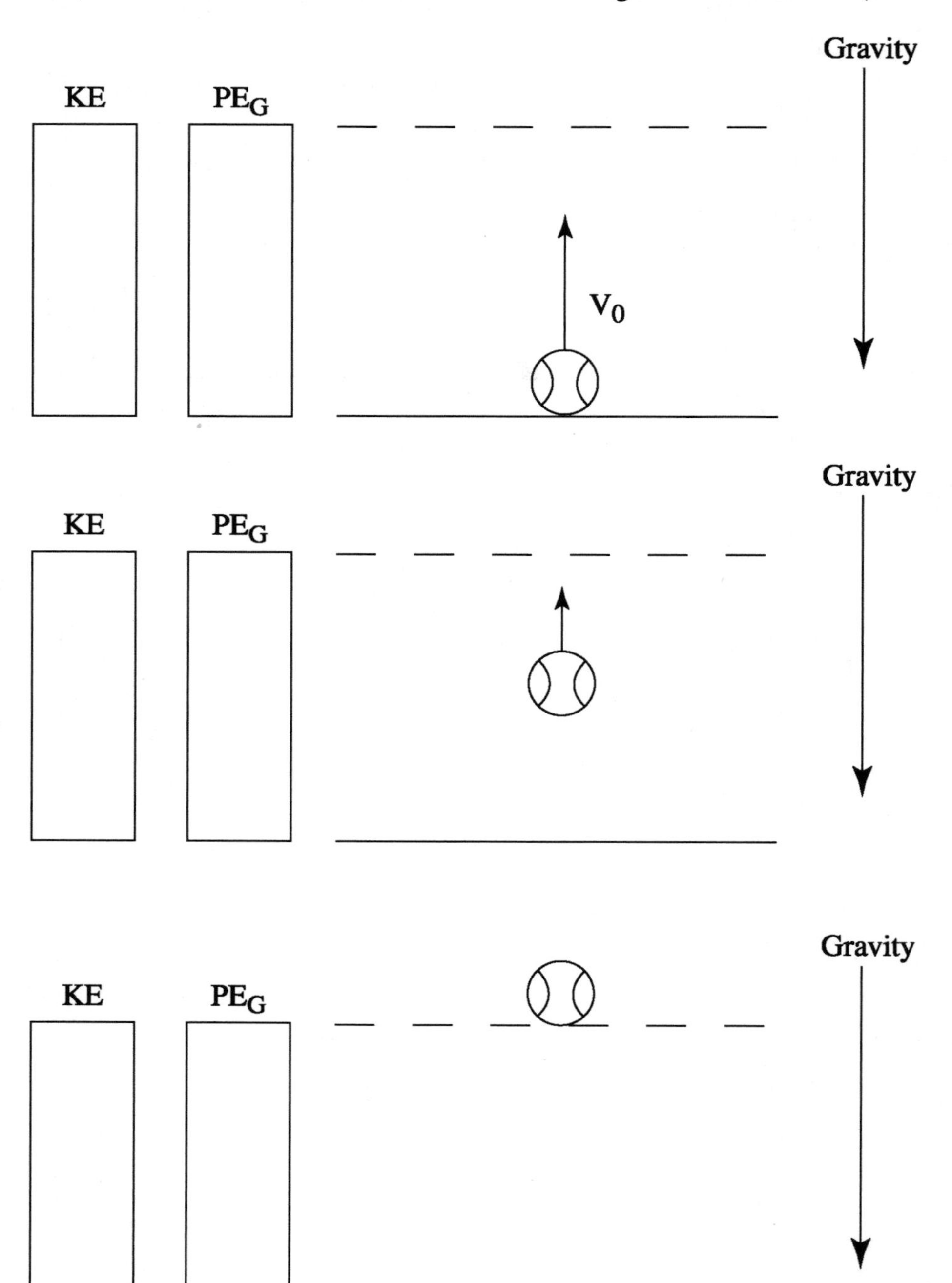

13. In the sketches seen on the following page, a positive charge is moving in a constant electric field and starts with an initial velocity of v_0. Fill in the energy bars as the charge moves and draw the relative velocity vectors for each frame. (The dotted line indicates where the charge changes directions.)

Name ______________________ Date __________ Class __________

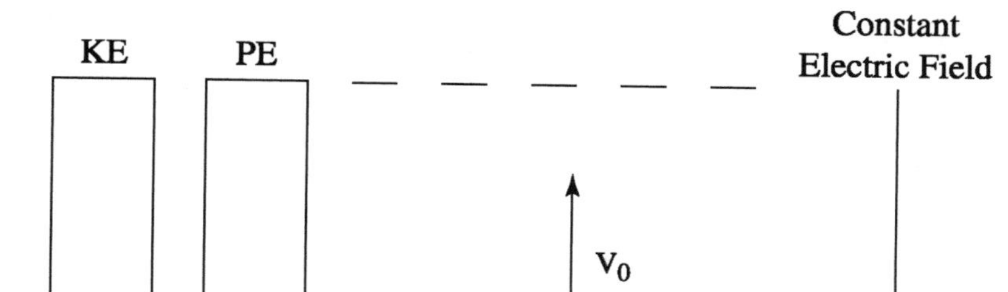

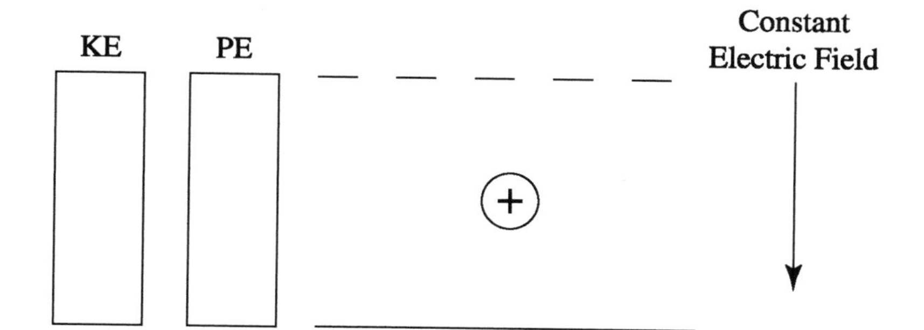

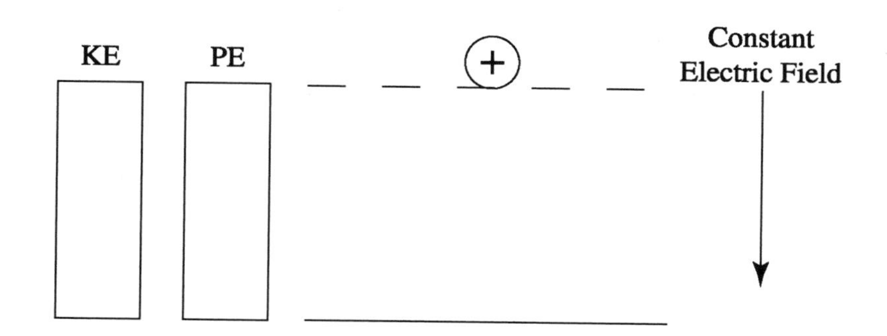

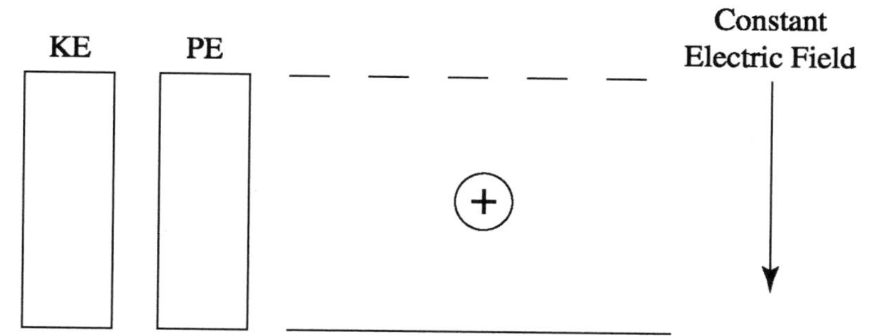

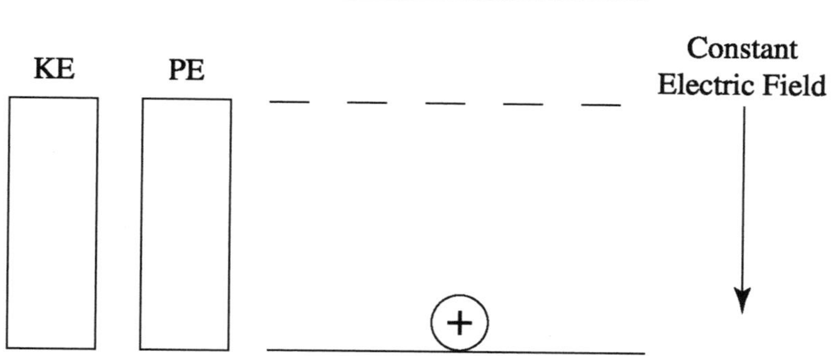

Name ______________________ Date ____________ Class ____________

Chapter 20

Quantitative Treatment of Current and Circuit Elements

1. Initially there are equal numbers of positive and negative charges on the plates of a capacitor. A battery is pushing an electron toward the left plate. (Both plates have a large equal number of positive and negative charges. Since there are an equal number of positive and negative charges, their net charge is zero and we will ignore their effects for the purposes of this problem.)

   a. Which end of the battery is the left plate connected to?

   b. Which end of the battery is the right plate connected to?

   c. The electron on the left-hand plate "pushes" on the electrons on the opposite plate. Draw all of the force vectors on the electron shown on the left-hand plate as well as the net force exerted on that electron. (Hint: What effect does the battery have on the electron?)

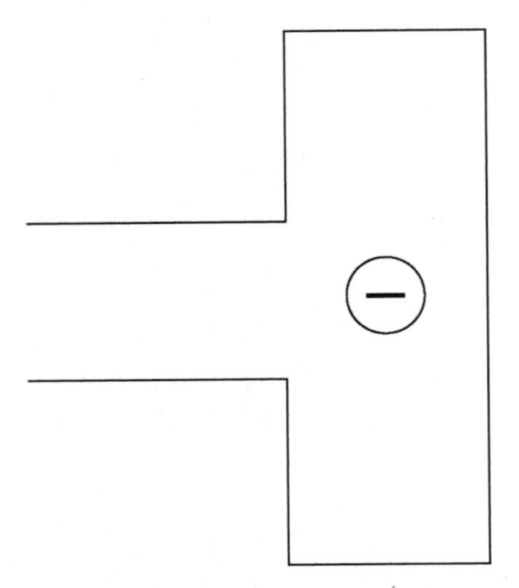

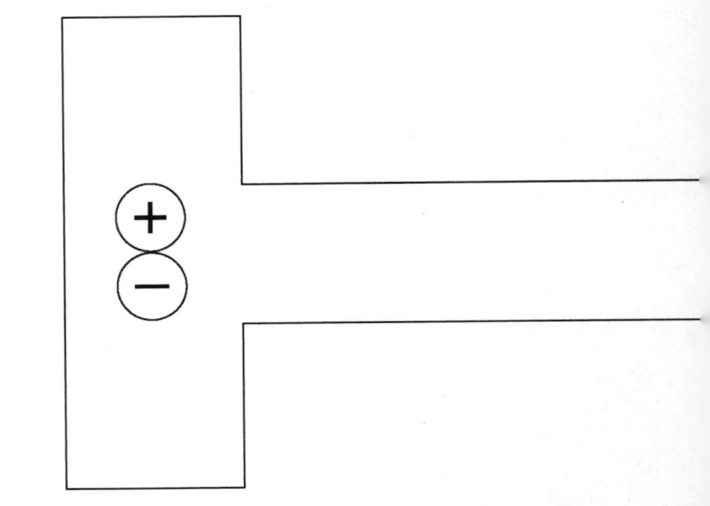

Name ________________ Date ________ Class ________

d. Draw the forces exerted on the electron on the right-hand plate as well as the net force.

e. As an electron moves through a wire toward the negative side of a parallel plate capacitor, it sees a large number of negative charges pushing it back and a large number of positive charges pulling it forward. Draw the forces exerted on a single electron as it approaches the capacitor. (Use one vector for the net effect of the negative charges and one vector for the net effect of the positive charges.)

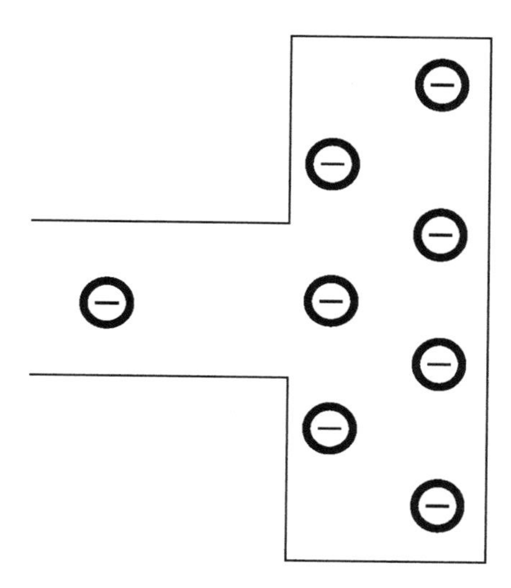

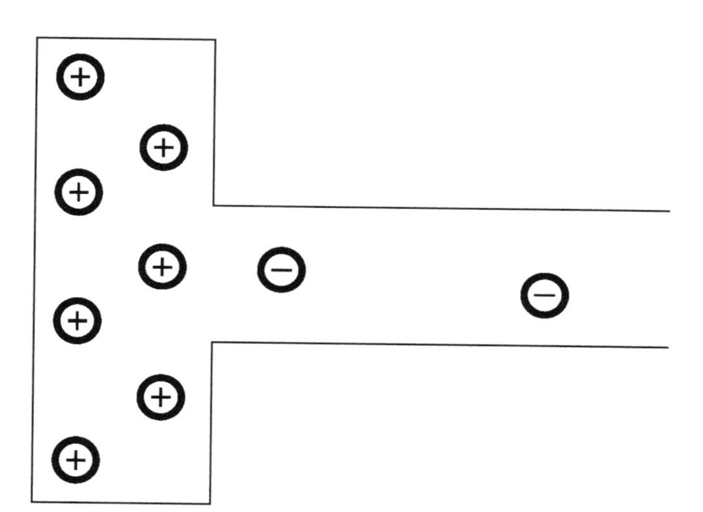

f. After some time has passed, there is even more of a build-up of net charge on both plates. Draw the force vectors on an electron approaching the negative plate. How do the forces on the electron compare with the forces in the previous problem?

g. Once the capacitor is fully charged, no more electrons can move onto the plate. Draw the forces exerted on an electron approaching the negative plate. Which direction is the net force exerted on the electron pointing?

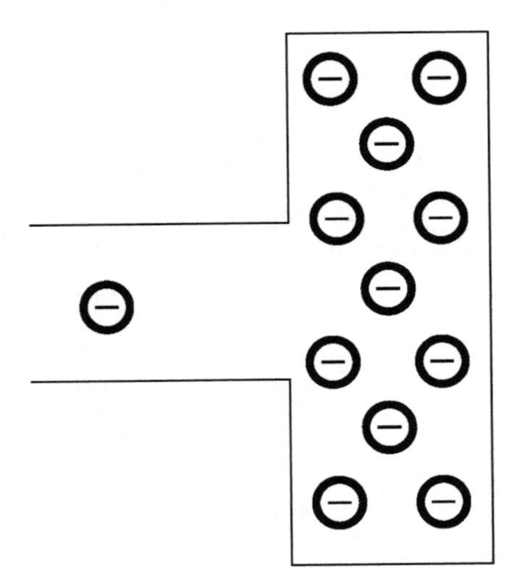

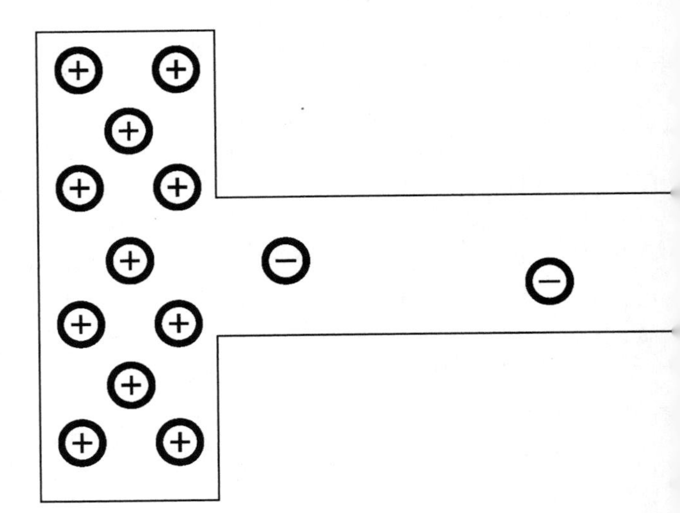

h. The plates of the capacitor from the previous problem are pushed closer together. Draw the forces experienced by the incoming electron in this case. Explain how the forces have changed from the previous problem.

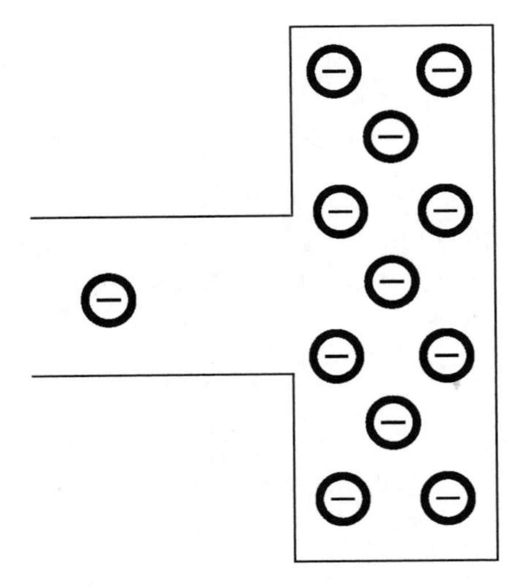

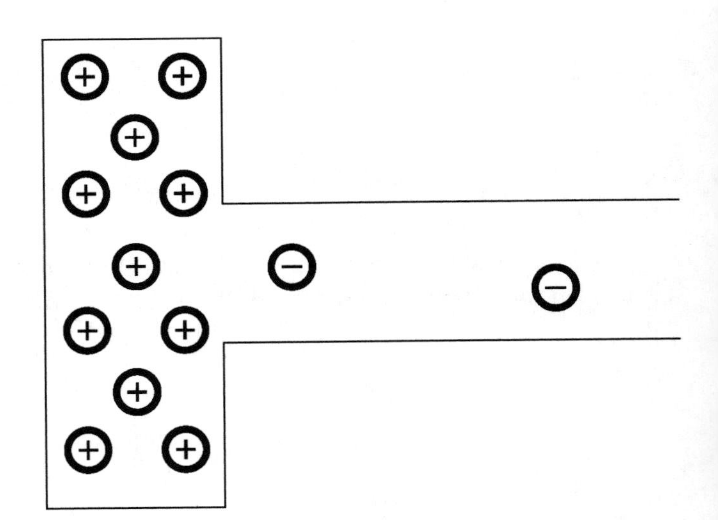

Name ______________________________ Date ____________ Class ____________

i. Starting from the full capacitor in problem ***g***, sketch the capacitor if the area of each plate were increased (e.g., increase the height of the plates). Assume the net charge on both plates is the same as in ***g***. Draw the force experienced by the incoming electron in this case and compare the net force to ***g***.

j. Help your friend study for a quiz by explaining what capacitance measures and how changing the distance between plates or changing the area of the plates affect capacitance.

2. You will develop an analogy of traffic on a road to describe how resistance in a wire depends on the diameter, length, and resistivity of the wire.

   a. The parameters you can change are the number of lanes of traffic, the length of the road, and the average number of pot holes in a section of the road (one lane wide). Match up which properties of a wire these three parameters are analogous to.

   b. In this analogy, what do the cars represent?

   c. What does the average speed of the cars represent?

   d. What does the number of cars leaving the length of road during each time period represent?

e. Sketch the following in the space provided:

   i. A short length of road, single lane, with only a few pot holes for each section of the road.
   ii. A short length of road, three lanes, with only a few pot holes for each section of the road.
   iii. A long length of road, one lane, with only a few pot holes for each section of the road.

f. Use the three sketches to write up an explanation for a friend who needs help understanding resistance.

3. Which has the lower resistance, a wire of length L and resistivity R, or a wire with half the length and twice the resistivity?

4. Which has the lower resistance, a wire of diameter D and resistivity R, or a wire with half the diameter and twice the resistivity?

Name ______________________ Date ____________ Class ____________

Chapter 21

Quantitative Circuit Reasoning

Voltmeters have a large resistance, while ammeters have a very small resistance. The problems explore how these instruments should be hooked up so they do not change any of the voltages or currents flowing through a circuit.

1. Calculate the current through and voltage across the 40Ω resistor.

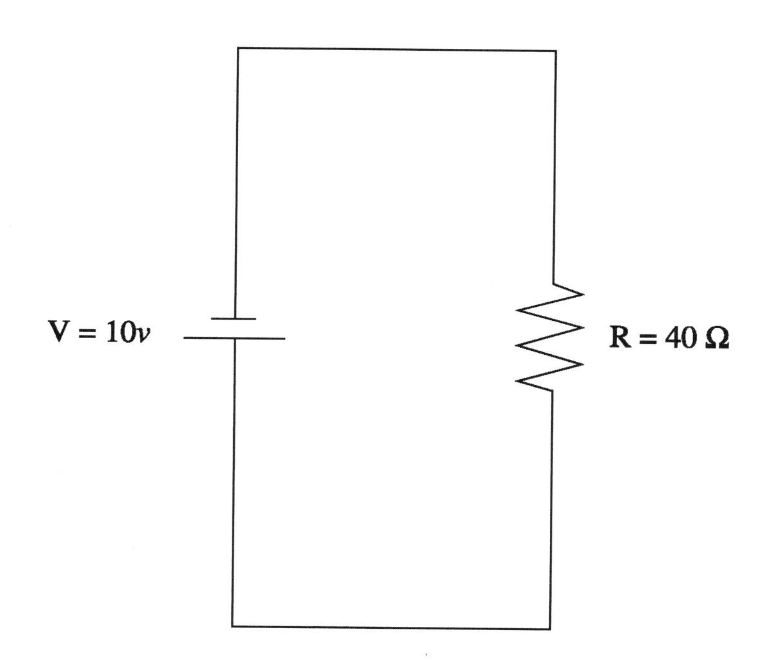

Name ______________________ Date ____________ Class ____________

2. Calculate the current and voltage for the 40Ω resistor when the voltmeter is connected as shown in the figure below. The resistance of the voltmeter is 10^6 Ω.

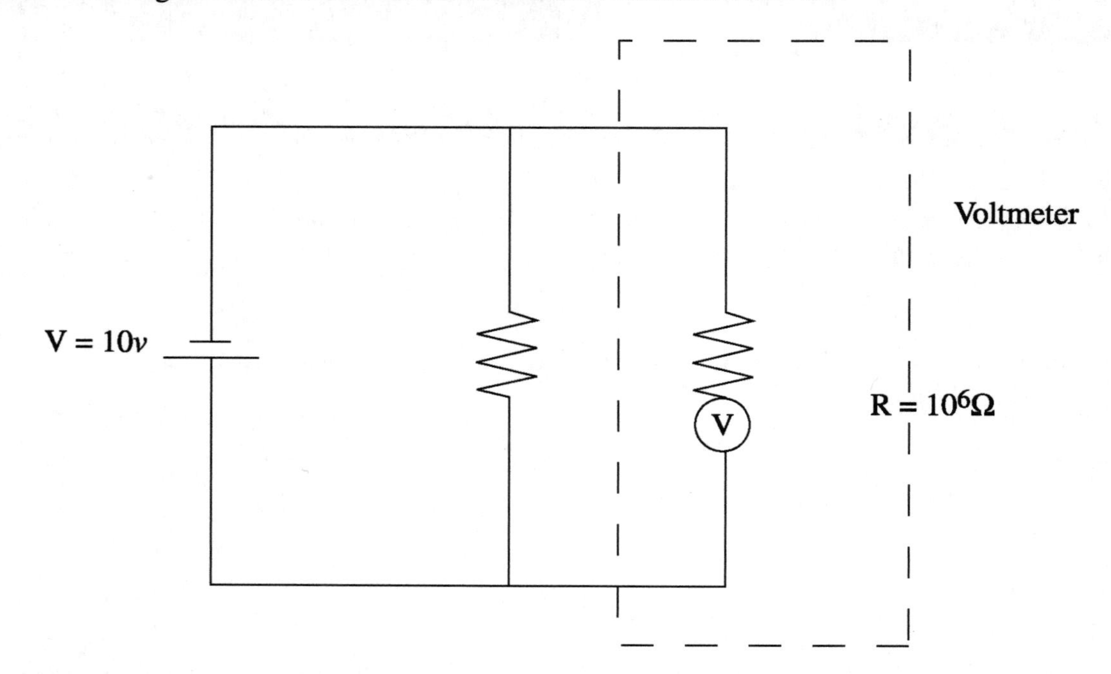

3. Calculate the current and voltage for the 40Ω resistor when the voltmeter is connected as shown in the figure below.

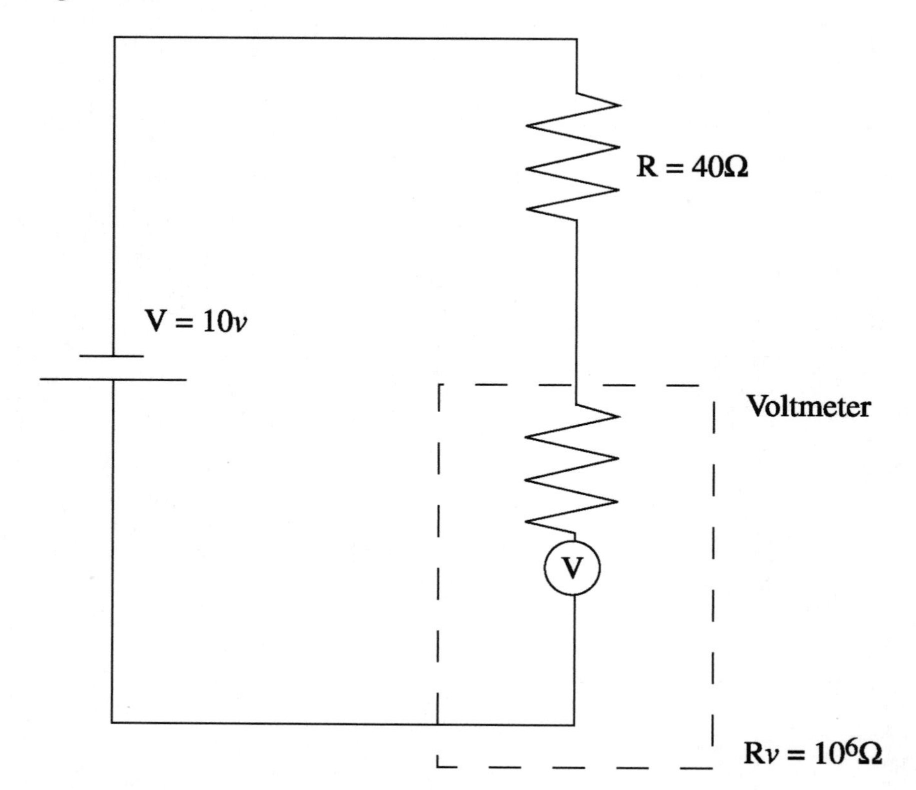

Name ______________________________ Date ____________ Class ____________

4. Which of the two configurations has the least effect on the current in the circuit?

5. What is the voltage drop across the 40Ω resistor for the configuration you chose in the previous problem?

6. In order to accurately measure the voltage drop across a resistor, how should you hook the voltmeter up?

7. Calculate the current and voltage for the 40Ω resistor when the ammeter is connected as shown in the figure below. The ammeter resistance is 0.001 Ω.

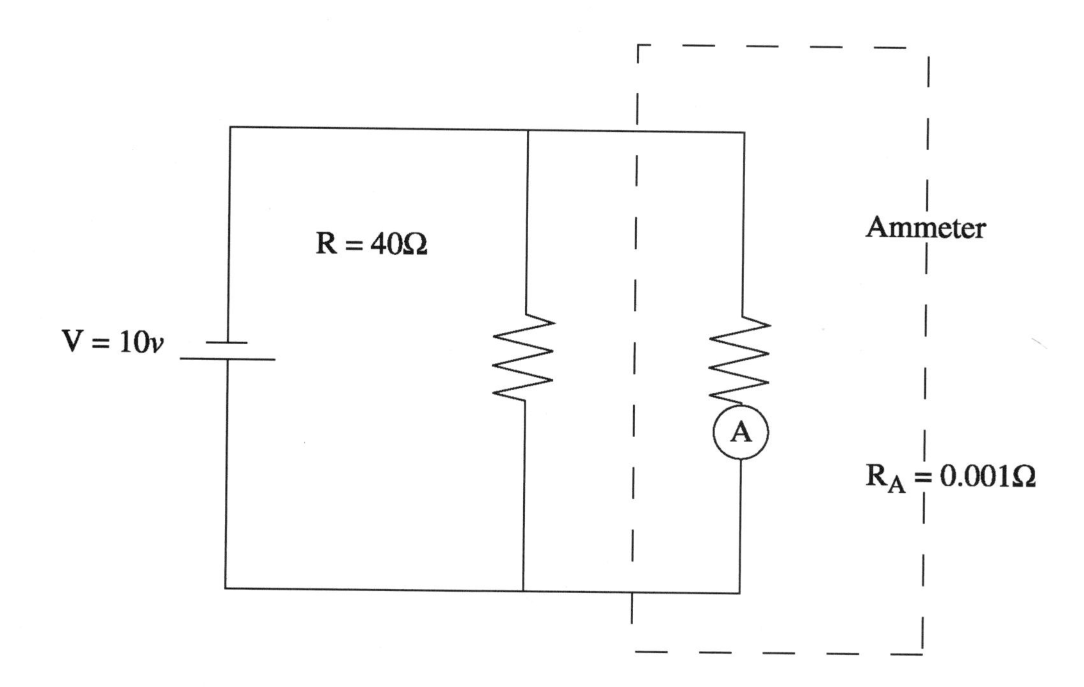

Name ______________________ Date ____________ Class ____________

8. Calculate the current and voltage for the 40Ω resistor when the ammeter is connected as shown in the figure below.

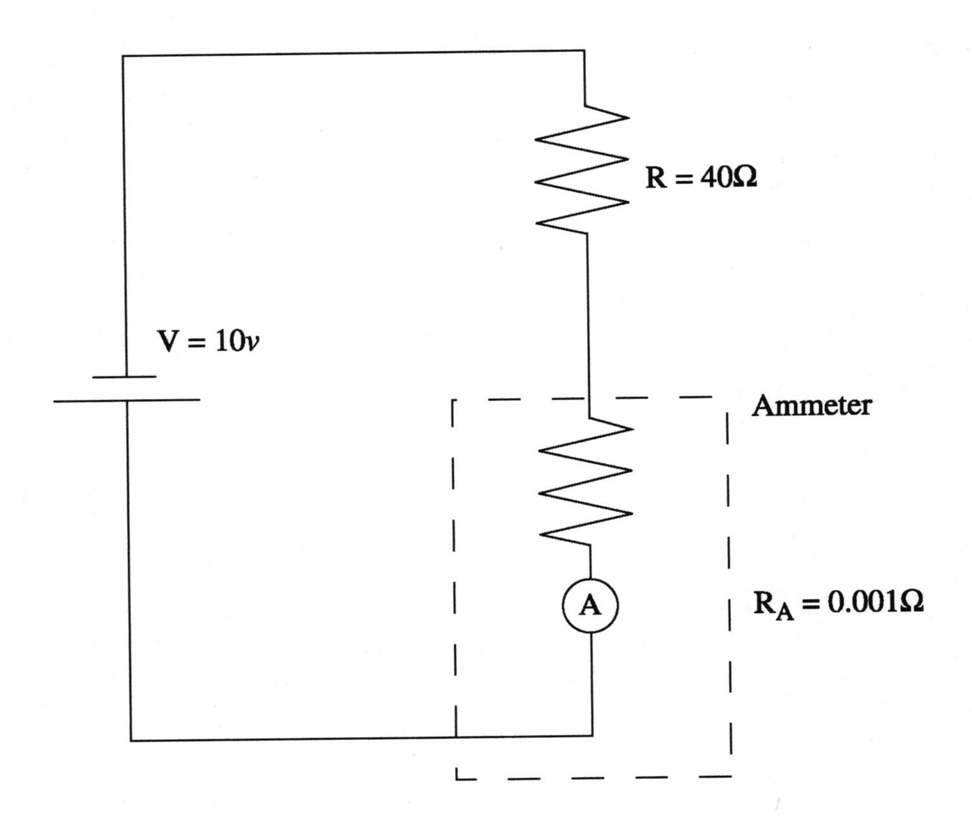

9. Which configuration changes the voltage and current in the original circuit the least?

Name ______________________ Date ____________ Class ____________

10. Calculate what a voltmeter would read if the leads were placed at the two points indicated.

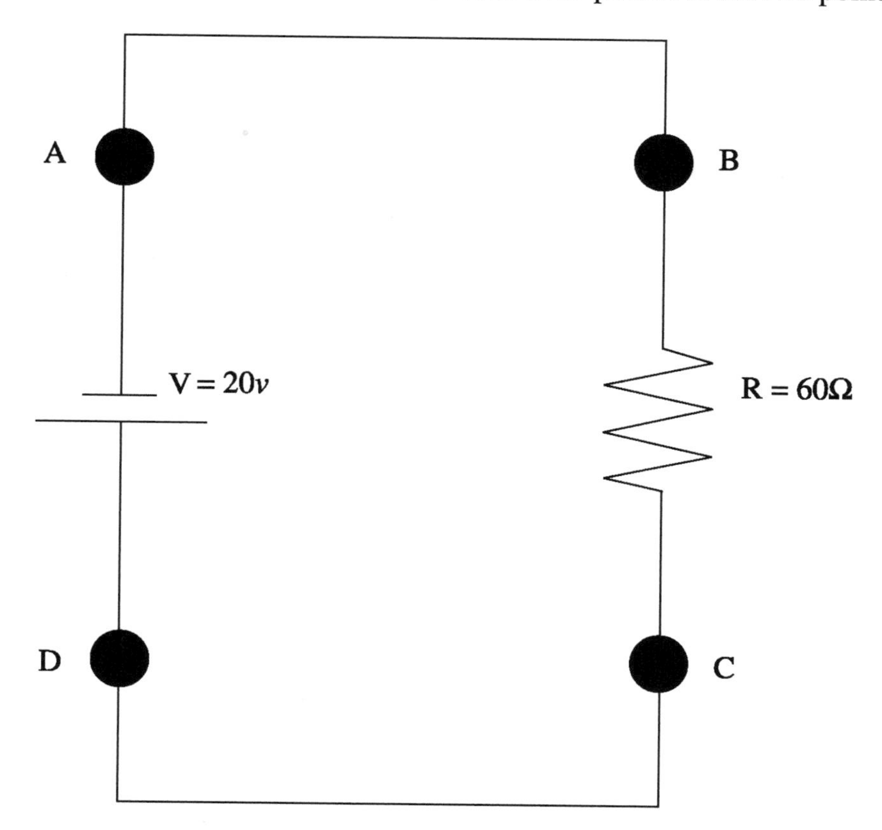

a. Potential difference between A and B

b. Potential difference between B and C

c. Potential difference between C and D

11. Calculate the reading an ammeter would show if connected in series at the following points.

a. Current measured at A

b. Current measured at B

c. Current measured at C

d. Current measured at D

Name ______________________ Date ____________ Class ____________

12. Calculate what a voltmeter would read if the leads were placed at the two points indicated.

a. Potential difference between A and B

b. Potential difference between B and C

c. Potential difference between C and D

d. Potential difference between B and D

13. Calculate the reading an ammeter would show if connected in series at the following points.

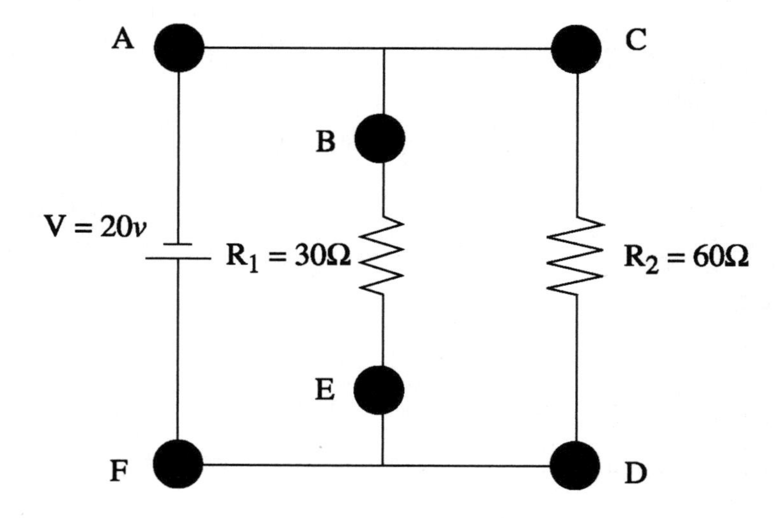

a. Current measured at A

b. Current measured at C

c. Current measured at E

Name ______________________________ Date ____________ Class ____________

14. Calculate what a voltmeter would read if the leads were placed at the two points indicated.

a. Potential difference between A and B

b. Potential difference between A and F

c. Potential difference between B and E

d. Potential difference between C and E

e. Potential difference between C and F

f. Potential difference between D and F

15. Calculate the reading an ammeter would show if connected in series at the following points.

a. Current measured at A

b. Current measured at B

c. Current measured at C

d. Current measured at D

e. Current measured at E

f. Current measured at F

Name ______________________ Date ____________ Class ____________

16. Calculate what a voltmeter would read if the leads were placed at the two points indicated.

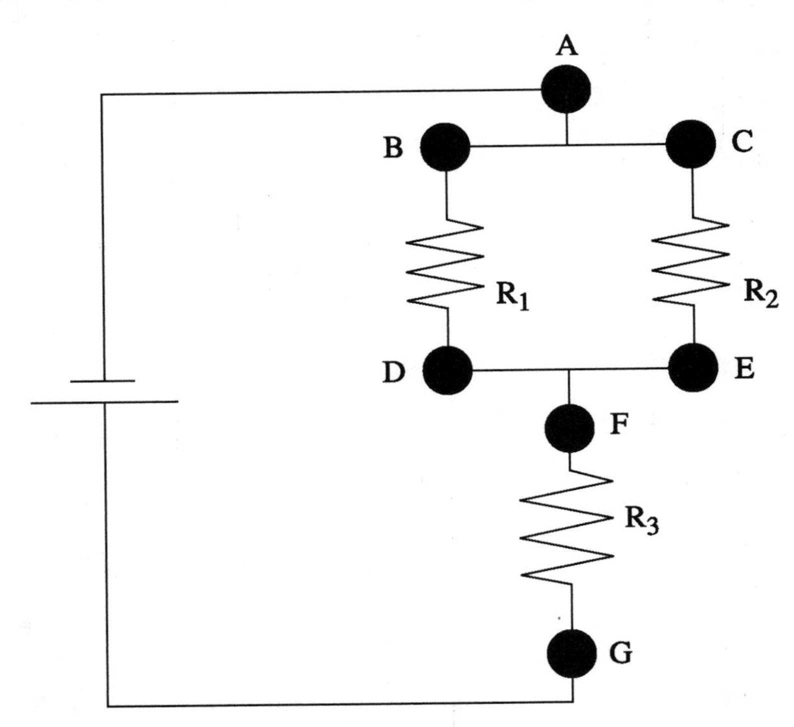

a. Potential difference between A and B

b. Potential difference between A and C

c. Potential difference between B and D

d. Potential difference between A and F

e. Potential difference between F and G

f. Potential difference between A and F

17. Calculate the reading an ammeter would show if connected in series at the following points.

a. Current measured at A

b. Current measured at B

Name ______________________ Date __________ Class __________

c. Current measured at C

d. Current measured at D

e. Current measured at E

f. Current measured at F

g. Current measured at G

18. Circle which 3 circuits are equivalent to each other. (The voltage drop across each resistor is the same for equivalent circuits.) Assume all resistors and batteries are identical.

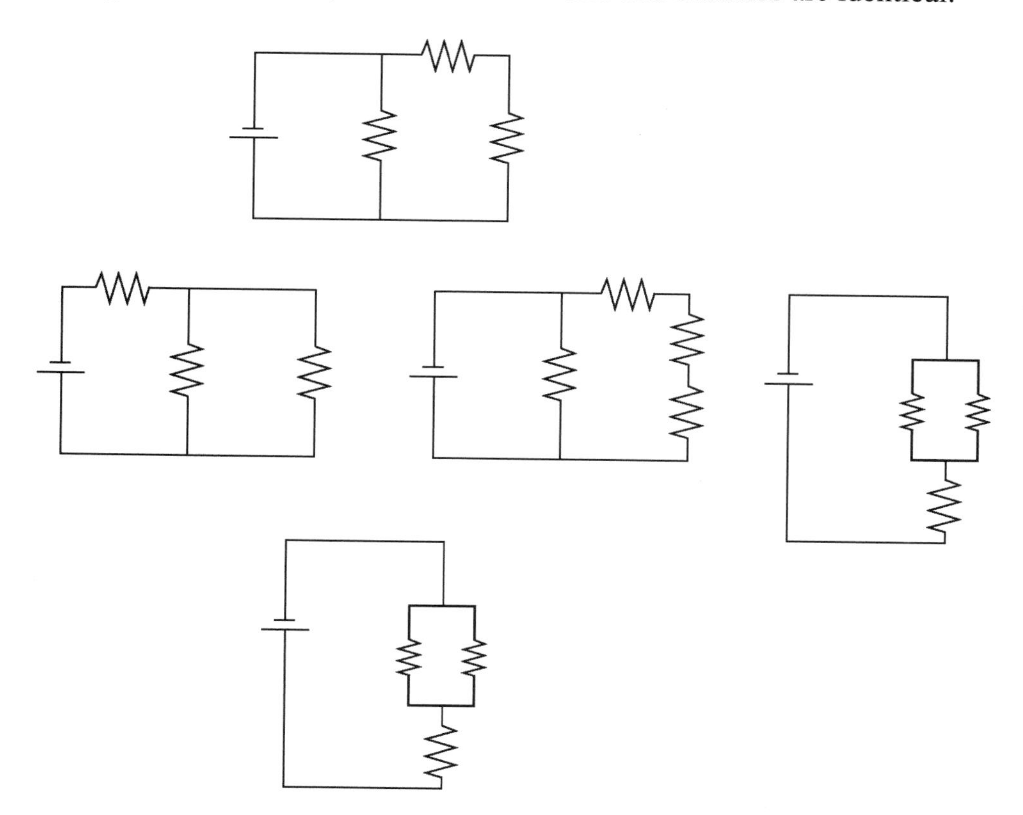

Name ______________________ Date __________ Class __________

19. Rank the current through the battery from largest to smallest current. Assume the resistors and batteries are identical.

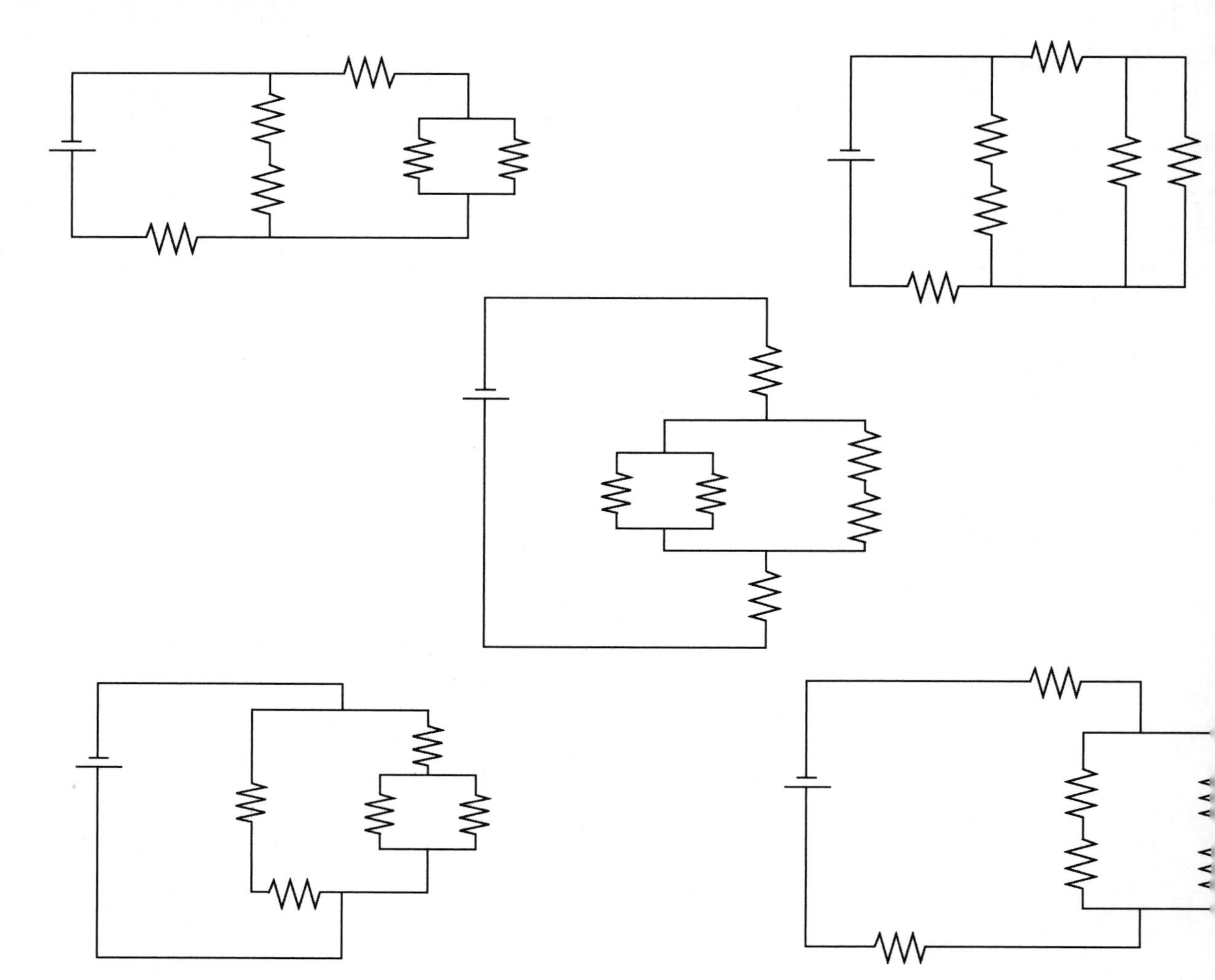

Name ______________________________ Date ____________ Class ____________

Chapter 22

Magnetism and Magnetic Fields

1. Draw the magnetic force vector on each item as each end of a bar magnet is brought close to the item.

   a. A stationary charge
   b. A paperclip
   c. Another magnet

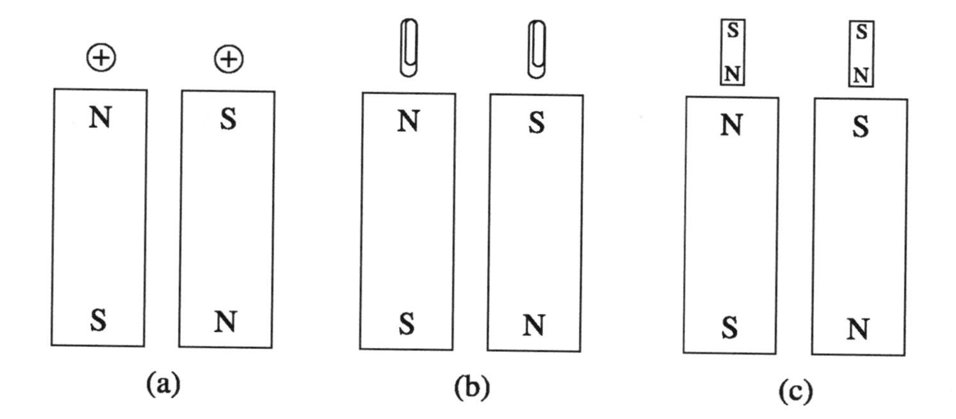

Name ______________________ Date ____________ Class ____________

2. Draw an arrow showing the direction the magnetic force is exerted on a current if the compass needle is as shown. Also show the direction of the magnetic field.

F	I	B	C
	→		N (top), S (bottom)
	↑		S (left), N (right)
	⊙		N (top), S (bottom)
	←		N (left), S (right)
	⊗		N (top), S (bottom)

Name ______________________________ Date ____________ Class ____________

3. Draw an arrow showing which direction the magnetic field must point given the direction of the current and the force exerted on the wire carrying the current. In the last column, draw a compass needle.

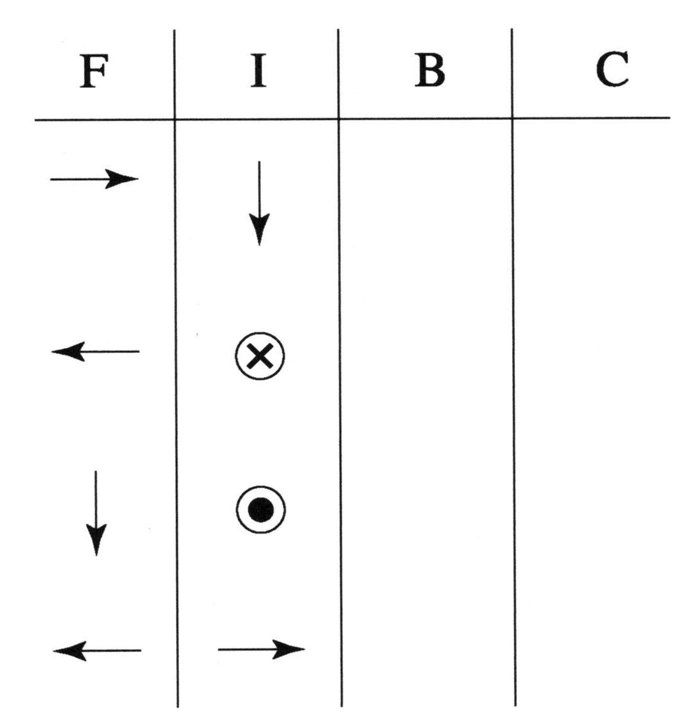

Name ______________________________ Date ____________ Class ____________

4. Draw the direction of current flow for the force on the wire indicated. (Assume currents run perpendicular to the magnetic field.) Also draw a compass needle in the last column.

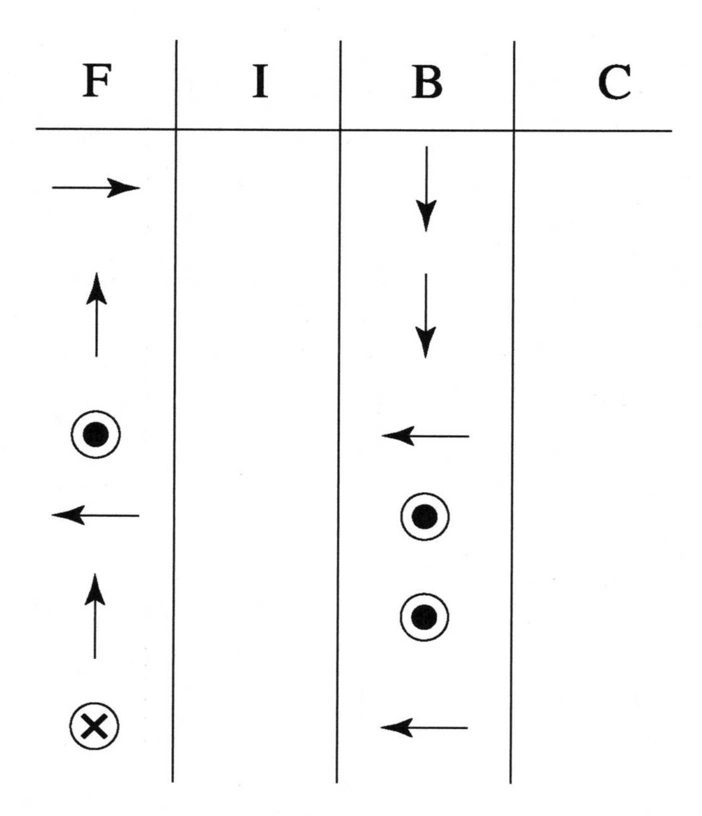

5. A rectangular coil of wire with a current flowing through it is oriented with its rotational axis perpendicular to the magnetic field (perpendicular to the surface of this page). The coil is stationary at t_1. (Assume any friction from the axle is negligible.)

   a. Does the torque on the coil change as it starts turning? If yes, explain why it changes.

   b. Will the coil continue turning in the same direction or will it reverse direction?

   c. At which time is the angular acceleration of the coil the greatest?

Name ______________________ Date ____________ Class ____________

d. At which time is the angular acceleration of the coil the least?

6. At t_1, the "width" of the coil is measured horizontally on the page and "length" is in to the page.

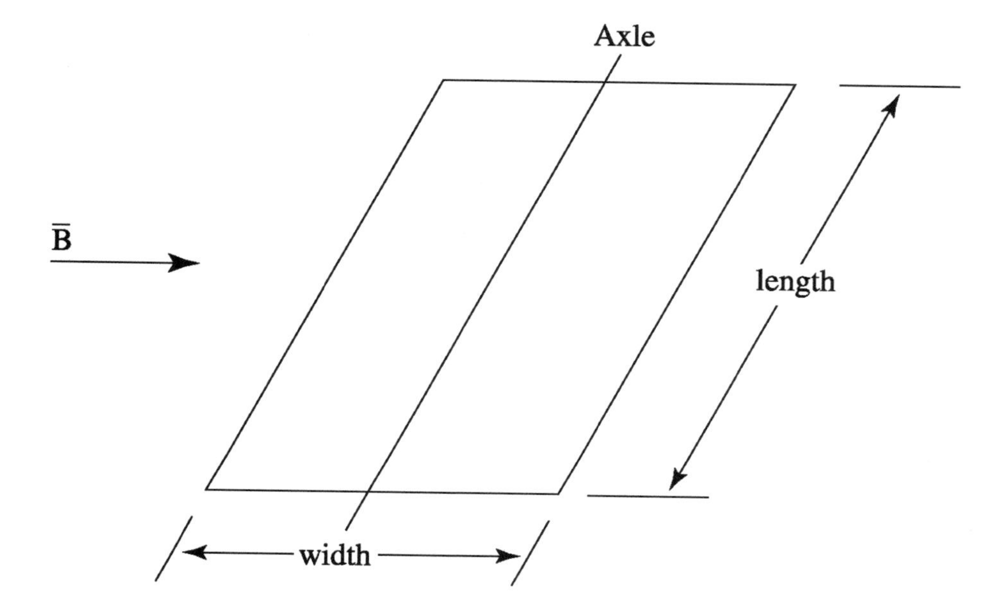

a. If the width of the coil is increased, does the maximum acceleration change?

b. If the width of the coil is increased, does the rate of oscillations change?

c. If the length of the coil is increased, does the maximum acceleration change?

d. If the length of the coil is increased, does the rate of oscillations change?

e. If the current through the coil is increased, does the maximum acceleration change?

f. If the current through the coil is increased, does the rate of oscillations change?

g. Can you increase the magnetic force on the coil without increasing the torque on the coil?

Name ______________________ Date ____________ Class ____________

Chapter 23

Electromagnetic Induction

1. A bar magnet is held close to a closed loop with the magnet aligned parallel to the axis of the coil. For each situation, determine whether the magnetic flux through the loop is increasing, decreasing, or unchanged.

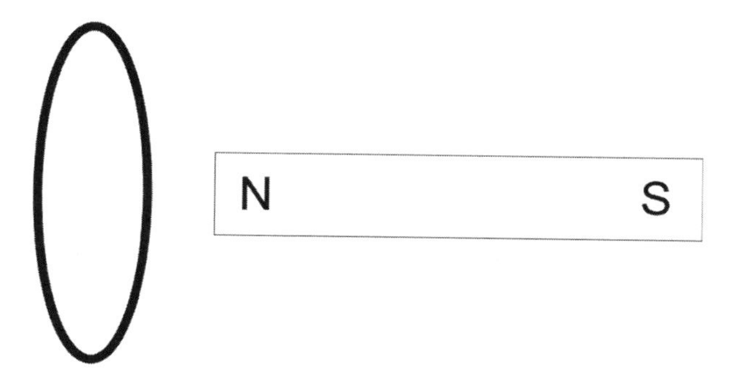

   a. The magnet moves toward the coil

   b. The magnet moves away from the coil

   c. The magnet rotates 90° (North end pointing up)

   d. The magnet rotates 180° (South end toward coil)

2. The figures below show a series of conducting objects moving through a uniform magnetic field. Indicate whether the magnetic flux enclosed is increasing, decreasing, or unchanged.

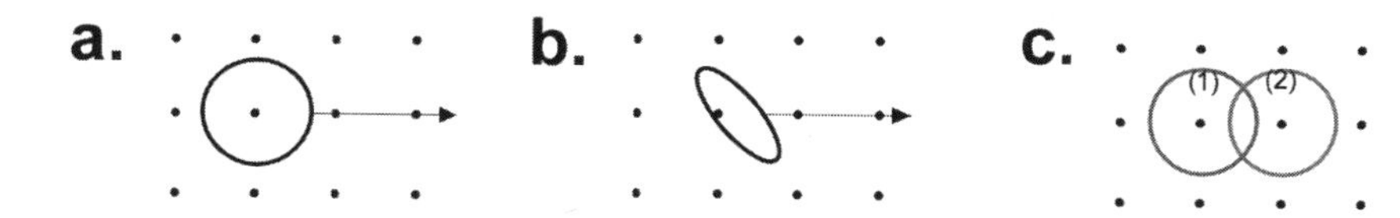

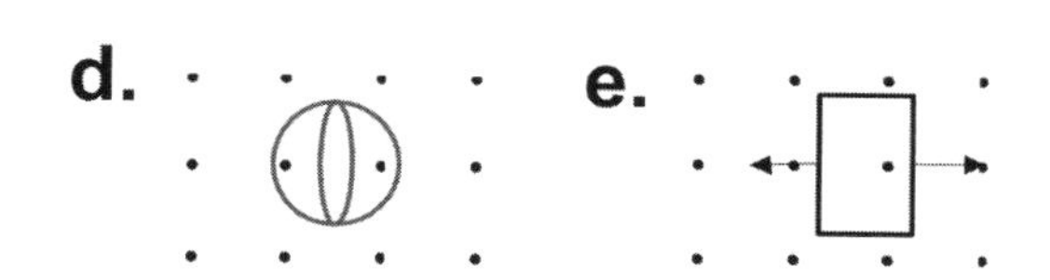

Name ______________________________ Date ____________ Class ____________

a. A ring moving toward the right at a constant velocity

b. An ellipse moving to the right at a constant velocity

c. A ring oscillating back and forth between positions (1) and (2)

d. A ring oscillating between a circle and an oval

e. A rectangular loop being stretched horizontally

3. A circular conducting coil is being stretched uniformly so the area enclosed by the loop is constantly increasing.

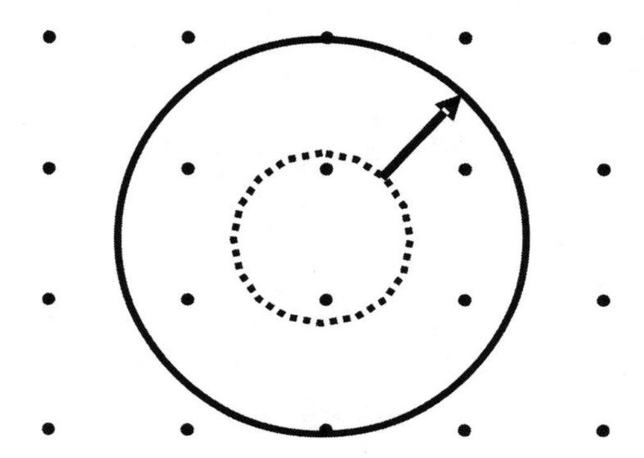

a. Is the magnetic flux enclosed by the loop increasing, decreasing, or unchanged?

b. In which direction is the induced current flowing?

c. If the magnetic field pointed into the page rather than out, what would happen to the magnetic flux and induced current as the loop expands?

4. Determine the direction of current flow for each situation (clockwise, counterclockwise, or zero).

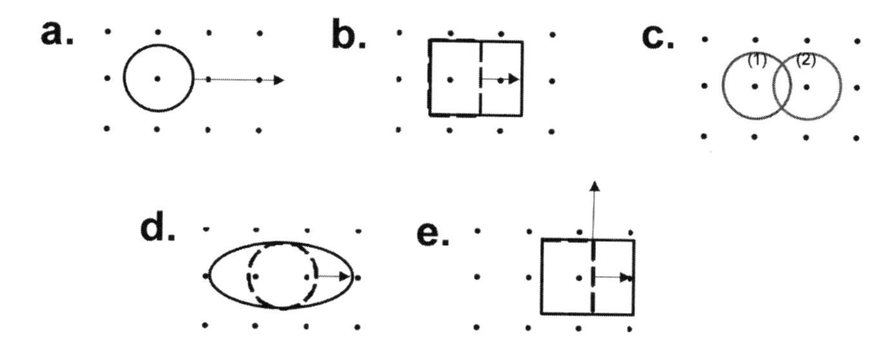

a. A ring moving toward the right at a constant velocity

b. A rectangular loop stretched towards the right

c. A ring oscillating back and forth between positions (1) and (2)

d. A ring stretching from a circle into an ever expanding oval

e. A rectangular loop being stretched to the right while moving at a constant velocity upward.

5. Rank, from smallest to largest, the induced current in a circular coil moving perpendicular to uniform magnetic field at constant velocity.

A Axis of coil oriented parallel to magnetic field moving with velocity v
B Axis of coil is 30° to magnetic field moving with velocity $3 \times v$
C Axis of coil is 45° to magnetic field moving with velocity $0.707 \times v$
D Axis of coil is 90° to magnetic field moving with velocity v
E Axis of coil is parallel to magnetic field and coil is stationary

Name ______________________ Date ____________ Class ____________

6. Two identical bar magnets are released from the same height at the same time. One magnet is dropped through a hollow metal tube and the other falls free of any objects.

   a. Draw a free body diagram for the magnet falling through the tube at four different points: a. before it enters the metal tube, b. as it starts to enter the tube, c. just as it starts to leave the tube, and d. after it has left the tube. Explain what is different between the four cases and why they are different.

a. Before magnet enters tube	**c.** As magnet leaves tube
● Explanation:	● Explanation:
b. Magnet enters tube	**d.** Magnet has left tube
● Explanation:	● Explanation:

Name ______________________________ Date ____________ Class ____________

b. Describe the velocity of the magnet falling through the tube compared to the freely falling magnet starting just before it enters the tube until just after it has fallen completely through the tube.

c. Can you determine which magnet hits the ground first? Explain your reasons for choosing your answer.

7. A rectangular conducting loop is dropped between two closely spaced poles of a "C" shaped electromagnet. Indicate whether the magnetic flux through the loop is increasing, decreasing, or constant in each of the situations described below. Additionally, indicate whether the current in the near side of the loop is up, down, or no current.

   a. Before the bottom of loop is between the magnet poles.

   b. Before the midpoint of the loop has fallen halfway between the poles.

   c. After the midpoint of the loop has fallen halfway between the poles.

   d. After the entire loop has passed between the poles.

8. A bar magnet is mounted on an air glider and the glider track passes through a closed loop coil.

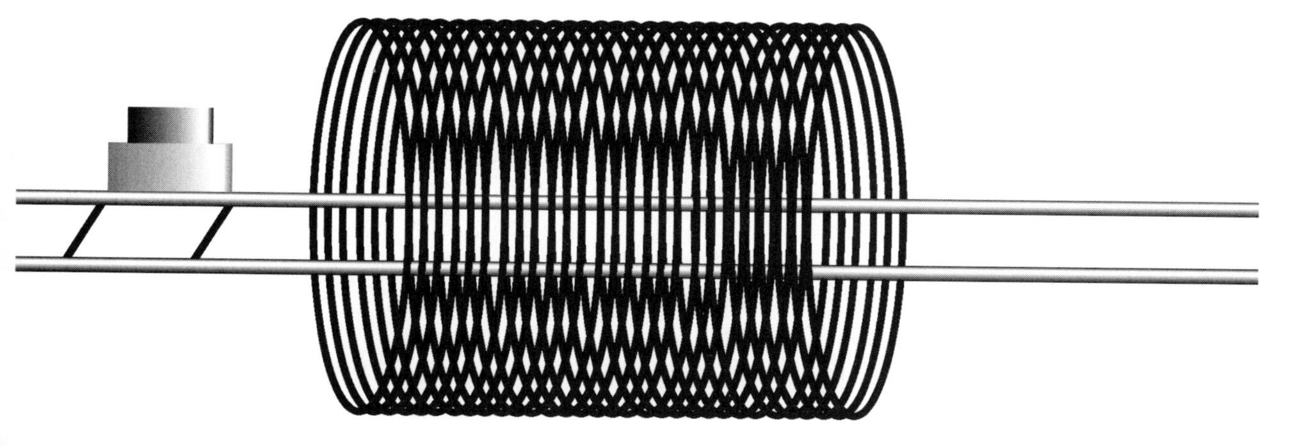

Name ______________________________ Date ____________ Class ____________

a. The glider is given an initial push. Sketch the position versus time graph from just before it enters the coil until just after leaving the coil.

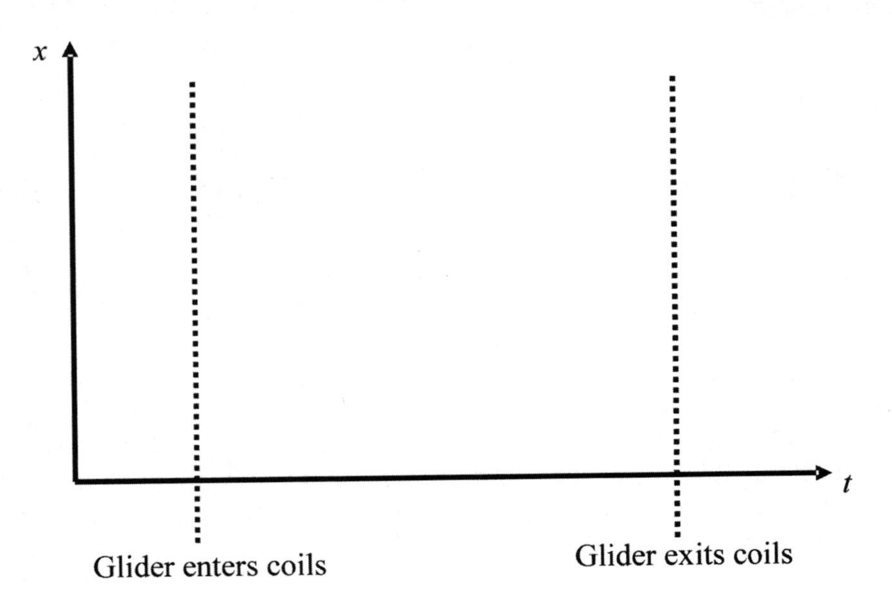

b. If the initial velocity is increased, how would the motion of the magnet change? Use a different color to sketch the position versus time graph above.

Name ______________________ Date __________ Class __________

Chapter 24

As the Twentieth Century Opens: The Unanswered Questions

1. Where is the electric field changing the quickest? (Compare how much the magnitude changes during one box of time on the graph below.)

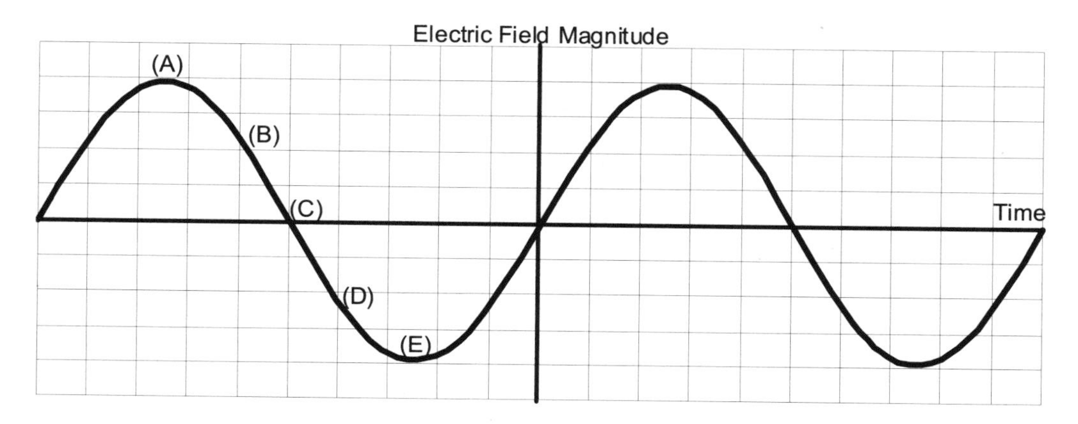

   a. Where is the electric field changing the least?

   b. Rank the rate at which the electric field magnitude is changing at points A through E, from least to most.

   c. The magnetic field is proportional to how quickly the electric field changes. Using a colored pen, sketch the magnetic field resulting from the changing electric field.

   d. Where is the magnetic field largest?

   e. Where is the magnetic field smallest?

Name ______________________________ Date ____________ Class ____________

2. Rank the rate at which the magnetic field magnitude is changing at points A through E, from least to most.

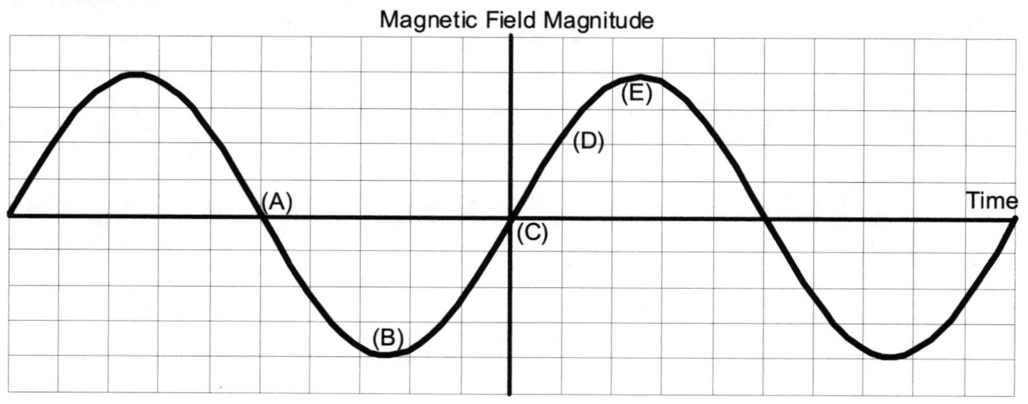

a. Sketch the electric field that results from the changing magnetic field.

b. When the electric field is at its maximum value, the magnetic field is _______.

c. When the electric field is zero, the magnetic field is ________.

d. The peak of an electric wave "lines up" with the _______ of a magnetic wave.

3. A sealed glass tube containing hydrogen has a glass window on one end and a diffraction grating for looking at the spectrum. If you used this tube to look at the light coming from a 1500°C stove burner, sketch what you would expect the spectrum to look like. Explain why you think the spectrum would look this way.

4. If you place the tube on the stove and let it heat up, what might the light emitted by the tube look like? Explain why you think the spectrum would look this way.

Name ______________________ Date ____________ Class ____________

5. A passenger on a bus is walking toward the front of the bus at 3 m/s. If the bus is crawling along at 20 m/s, how fast does a person on the curb see the passenger go by?

6. A driver moving 15 m/s in the opposite direction observes the walking passenger. How fast does driver think the passenger is moving relative to the car?

7. A northbound bus traveling 25 m/s passes a southbound bus going 31 m/s. How fast would a passenger on the northbound bus need to run so someone on the other bus would be stationary relative to the passenger?

8. How fast would a passenger on the southbound bus need to run so someone on the northbound bus is stationary relative to them?

Name ______________________________ Date ____________ Class ____________

Chapter 25

Relativity

1. Identify whether each reference system is inertial or noninertial.

   a. A train car on a straight, level track moving at a constant velocity.

   b. A rocket while it is being launched from the Earth's surface.

   c. A train car moving at a constant speed while it is rounding a corner.

   d. A fast moving car as it reaches the crest of a hill.

   e. A car stopped at a traffic light.

2. The Earth is, strictly speaking, not an inertial reference frame. Why not?

3. A passenger on a 747 traveling at 500 mph sees a jet approaching along a parallel path.

   a. If the jet is traveling at 450 mph, how fast does the passenger see the jet approaching?

   b. The pilot of the jet sees a flight attendant on the 747 running towards the front of the plane at 15 mph. How fast is she moving relative to the pilot of the jet?

   c. A baseball player on the 747 throws a baseball toward the back of the plane at 90 mph. How fast is the ball moving relative to the jet pilot?

Name ______________________________ Date ____________ Class ____________

4. A mail carrier on the sidewalk sees a bus moving at a constant velocity toward her. A passenger on the bus tosses his keys straight up in the air. Sketch the trajectory of the keys from the passenger's reference frame and the mail carrier's reference frame.

Trajectory of keys as seen by passenger	Trajectory of keys as seen by mail carrier

5. A motorcycle messenger sees the key toss as she passes the bus, going slightly faster than the bus. A taxi driver traveling the opposite direction as the bus also sees the passenger toss the keys. Sketch the trajectory of the keys for the messenger and taxi driver.

Trajectory of keys as seen by messenger	Trajectory of keys as seen by taxi driver

Name ______________________________ Date ____________ Class ____________

6. Anna and Adrian are both standing on a 100 m long barge. At either end of the barge are flashing bulbs, and Anna, located in the middle of the barge, sees the bulbs flashing together. Adrian is located halfway between Anna and bulb #1.

   a. Calculate the time it takes for light to travel from bulb #1 and from bulb #2 to Anna.

   b. Calculate the time it takes for light to travel from each bulb to Adrian.

   c. Does Adrian see both bulbs flashing at the same time? Which bulb does he see first?

   d. Anna sees bulb #1 flashing once each second. What is the time between flashes from Adrian's point of view?

7. Bonnie and Bill are on a second barge moving at relativistic speed past Anna and Adrian. Bill and Bonnie are on opposite ends of a longer barge.

   a. Just as Bonnie pulls even with bulb #2, it flashes. Does Bonnie see bulb #2 flashing at the same time?

   b. Write down the order in which Bill sees the bulbs flash.

   c. Just as Bonnie pulls even with Anna, she sees bulb #1 flash. Does Bonnie see bulb #2 flash at the same time? Explain why or why not.

   d. Bonnie pulls up even with bulb #1 when it flashes. Write down the order in which she sees the bulb flash.

e. When Bonnie is even with bulb #1, Bill still hasn't passed bulb #2. Write down the order Bill sees the lights flash.

f. Write down the order Bill sees the lights flash when he pulls even with Anna. Also write down the order in which Bonnie sees the bulbs flash.

g. Once they have moved past Adrian and Anna, what order do Bonnie and Bill see the lights flashing?

8. A passenger on a train throws a ball into the air and catches it 0.4 s later. A conductor on another train sees this occur. The two trains are moving at 0.5c relative to one another.

   a. How high does the ball go, according to the passenger?

   b. How high does the ball go, according to the conductor?

   c. How long does the conductor think the ball is in the air?

9. Calculate the fraction of *c* for the velocity and the time dilation for a proper time of 1 hour for each of the relative velocities.

Velocity	**v/c**	**Time**
100 m/s		
1000 m/s		
10^6 m/s		
10^7 m/s		
10^8 m/s		

Name ____________________ Date ____________ Class ____________

10. Calculate the fraction of *c* for the velocity and the length of a meter stick for the relative velocities.

Velocity	**v/c**	**Length**
100 m/s		
1000 m/s		
10^6 m/s		
10^7 m/s		
10^8 m/s		

Name ______________________ Date ____________ Class ____________

Chapter 26

Inroads into the Micro-Universe of Atoms

1. A group of tiny balls are shot at a target hidden under the rectangle in the diagram below. The balls bounce off a solid target. The target is a simple geometric shape. Based on the rebounding balls, sketch the shape of the target. Be sure to justify your sketch.

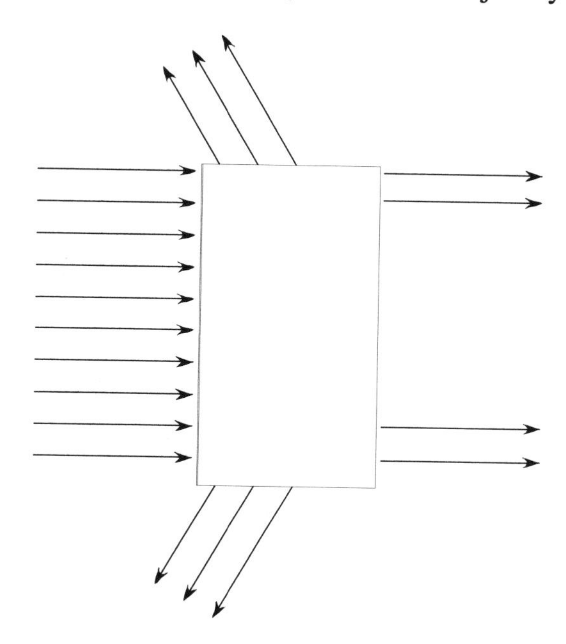

Name ______________________________ Date ____________ Class ____________

2. A group of tiny balls are shot at a target hidden under the rectangle in the diagram below. The balls bounce off a solid target. The target is a simple geometric shape. Based on the rebounding balls, sketch the shape of the target. Be sure to justify your sketch.

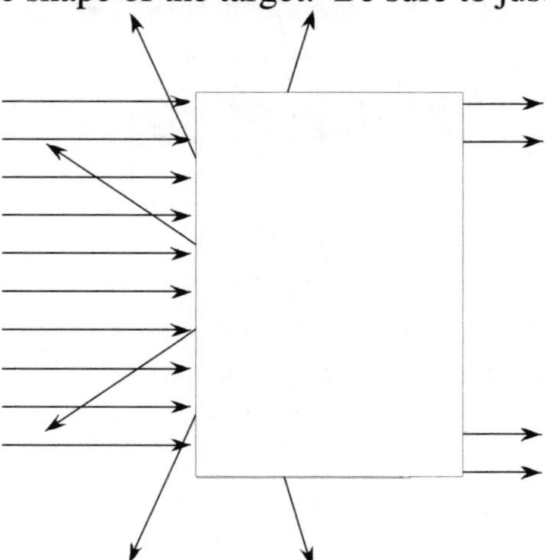

3. If the incident balls were traveling at a slower speed, how would path of the rebounded balls change? Sketch the path you think the balls would travel in this case.

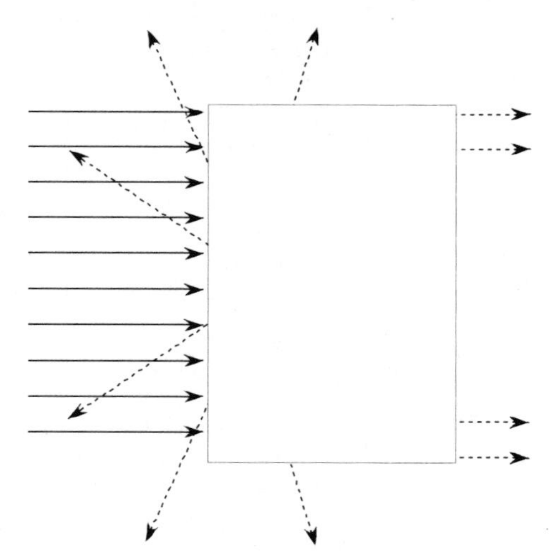

Name ______________________ Date ____________ Class ____________

4. When ultraviolet light shines on a negatively-charged electroscope, the leaves fall, while for a positively-charged electroscope, there is no observable effect. Based on these two facts, what is happening to the electroscope?

5. When ultraviolet light shines on a negatively-charged electroscope, the leaves fall, while when shined on a neutral electroscope, the leaves do not rise at all. Why doesn't the neutral electroscope gain a net charge from the photoelectric effect?

6. Why does classical physics predict that for low intensity light there can be a delay between when the light is turned on and electrons are detected?

7. Why was it so surprising to physicists when photoelectric current had a threshold below which no current flows?

8. Calculate the first four wavelengths of light in the Paschen series ($m = 3$, $n = 4, 5, 6, 7$).

Name ______________________________ Date ______________ Class ______________

Chapter 27

The Concepts of Quantization

1. What does the stopping voltage tell you about electrons? (Not the definition of stopping voltage, but rather, what information you can glean from the stopping voltage.)

2. Calculate the energy in joules and electron voltages for a photon with the following frequencies:

   a. $1.8 \times 10^{14}\ s^{-1}$

   b. $5.3 \times 10^{14}\ s^{-1}$

   c. $9.7 \times 10^{14}\ s^{-1}$

   d. $2.0 \times 10^{13}\ s^{-1}$

3. The total energy for an atom is:

$$E_n = -\frac{1}{n^2} E_1$$

   What does it mean for an atom to have a negative energy?

Name ______________________________ Date ____________ Class ____________

4. The energy equation and Rydberg equation are:

$$E_n = -\frac{1}{n^2}E_1 \text{ and } \frac{1}{\lambda} = R_H\left(\frac{1}{m^2} - \frac{1}{n^2}\right)$$

Is it a coincidence that these two equations appear so similar? What is the relationship between these two equations?

5. The energy equation is for the total energy of the atom. What types of energy contribute to the whole?

6. Classical physics does not predict the energy and radius of an atom are quantized. At large n values, does quantum theory predict nearly continuous radii and energies?

   a. Calculate R_n and E_n for a hydrogen atom with $n = 100$ and $n = 101$.

   b. Calculate R_n and E_n for $n = 1000$ and $n = 1001$.

   c. Compare the spacing of the energy and radii between the $n = 100$ and $n = 101$ to the spacing between $n = 1000$, and $n = 1001$.

   d. As n gets larger and larger, how do the quantum results agree with the classical results?

Name ______________________________ Date ____________ Class ____________

7. One consequence of relativity is that distances and time are not independent of one another. What comparable properties are not independent of each other?

8. Calculate the uncertainty in momentum for a hydrogen particle confined to a box (i.e., you know the atom is located somewhere within the box dimensions) of the size given below. Also calculate the uncertainty in the particles velocity.

   a. Hydrogen atom in a 1 mm box

   b. Hydrogen atom in a 1 μm box

   c. Hydrogen atom in a 1 Å box

   d. Hydrogen atom in a 0.1 Å box

   e. Calculate the average thermal velocity for a hydrogen atom at 300K. When does the uncertainty in the velocity become comparable to the average thermal velocity?

9. Why don't we experience quantum effects in everyday life?

Name ______________________ Date ____________ Class ____________

Chapter 28

The Nucleus and Energy Technologies

1. A friend is explaining how nuclear decay occurs. He states that Carbon-11 has a half-life of roughly 20 minutes. He then tells you that if you start with one million atoms, after 10 minutes 75 000 atoms will remain, after 20 minutes only half remain, and after 40 minutes all of the atoms will have decayed.

   a. Explain why your friend is or is not correct.

   b. How many Carbon-11 atoms are left after 10 minutes?

   c. How many Carbon-11 atoms are left after 40 minutes?

2. What fraction of atoms remains after one third of a half-life has gone by?

3. What fraction remains after one half of a half-life has gone by?

Name ______________________ Date ____________ Class ____________

4. Element 114 only has a half-life of 30 s. Fill in the table below with the average number of atoms left after the given time if you start off with 10 atoms.

Elapsed Time	**Atoms Left**
0 s	10
30 s	
60 s	
90 s	
120 s	
150 s	

a. On average, how long will it take for all 10 atoms to decay?

5. Using Equation 28-2, it would seem that it would take an infinite length of time for all atoms in a sample to decay. Obviously this is not true. What fact does this equation not take into account?

6. What does half-life mean for a single atom?

7. Supply the element or atomic number for these atoms.

a. ${}^{14}_{6}X$

$X =$

b. ${}^{24}_{Z}Mg$

$Z =$

c. ${}^{93}_{41}X$

$X =$

d. ${}^{27}_{Z}Al$

$Z =$

Name ______________________ Date ____________ Class ____________

e. Given only the element and the atomic number, why can't you uniquely determine the nucleon number? (For example ${}^{A}_{9}F$.)

8. What are the three conservation rules that nuclear reactions obey? Give an example for each rule.

9. Determine whether each nuclear reaction below violates charge conservation, baryon conservation, or both.

a. ${}^{210}_{84}Po \rightarrow {}^{204}_{82}Pb + {}^{4}_{2}He$

b. ${}^{238}_{92}U \rightarrow {}^{233}_{90}Th + {}^{4}_{2}He$

c. ${}^{11}_{6}C \rightarrow {}^{11}_{5}B + {}^{1}_{1}H$

d. ${}^{239}_{93}Np \rightarrow {}^{238}_{94}Pu + {}^{0}_{-1}e$

10. A helium ${}^{4}_{2}He$ atom is comprised of two protons, two electrons, and two neutrons. Add up the masses of the particles and compare that to the mass of the helium atom. Explain why these two values are different.

Name ______________________________ Date ____________ Class ____________

11. The mass of a $^{4}_{2}He$ helium atom is 4.003 u, while the mass of an $^{16}_{8}O$ atom is 15.999 u. The $^{16}_{8}O$ atom has four times as many protons, neutrons, and electrons as a $^{4}_{2}He$ atom. Explain why the mass of an $^{16}_{8}O$ atom is less than four times the mass of a $^{4}_{2}He$ atom.

12. Two hydrogen atoms can undergo fusion to form helium under certain conditions.

a. Why don't two hydrogen atoms in, say, a hydrogen balloon undergo fusion?

b. What prevents this from happening?

c. What role does the nuclear strong force play during fusion?

d. The roles of the electrostatic and strong force are reversed for fission. Explain what roles the two forces play and why they are reversed for fission.

✍ **Notes!**

✍ **Notes!**

✍ Notes!

✍ Notes!

✍ Notes!

✍ **Notes!**

✍ **Notes!**

✍ Notes!

✍ **Notes!**

✍ Notes!